Digital Filters
For Everyone
Third Edition

Rusty Allred

⌂ Creative Arts & Sciences House
West Melbourne, Florida USA

Copyright © 2015 by Creative Arts & Sciences House.
All rights reserved.

Creative Arts & Sciences House
West Melbourne, Florida USA

ISBN 978-0-9829729-2-2

Current Printing: 1

Printed in the United States of America

Library of Congress Control Number: 2014921339

Contents

Part 1: Introduction & Theory 1
  Chapter 1: Digital Filter Theory 1
    1.1.1 Basic System Theory 1
    1.1.2 Mathematical Review 4
    1.1.3 Definitions 5
    1.1.4 Basic Filter Theory 7
    1.1.5 The Transfer Function 9
    1.1.6 Filter Structures 13
    1.1.7 Filter Analysis 17
    1.1.8 Filter Transforms 42
    1.1.9 Filter Theory Summary and Quick Reference 49
  Chapter 2: Digital Filter Applications 53

Part 2: The Handbook 69
  Chapter 3: Infinite Impulse Response (IIR) Filters 69
    2.3.1 Butterworth Filters 69
    2.3.2 Linkwitz-Reilly Filters [Reference 17] 80
    2.3.3 Bessel Filters [Reference 5] 86
    2.3.4 Chebychev Type I Filters 94
    2.3.5 Chebychev Type II Filters 106
    2.3.6 Variable Q Filters 116
    2.3.7 Allpass Filters [References 8, 26, 27, 34] 122
    2.3.8 Equalization Filters [References 8, 23, 26, 27] 124
    2.3.9 Notch Filters [Reference 27] 126
    2.3.10 Shelf Filters [References 20, 21, 26] 127
    2.3.11 Troubleshooting the Filter Formulas 129
    2.3.12 Fun with Poles and Zeros 129
    2.3.13 After Note 133
  Chapter 4: Finite Impulse Response Filters 135
    2.4.1 Linear Phase FIR Filters 135
    2.4.2 Designing FIR Filters: Fourier Series Method 135
    2.4.3 Designing FIR Filters: Frequency Sampling Method 154
    2.4.4 Fun with Zeros 162
  Chapter 5: Two Dimensional Filters 167
    2.5.1 Some Standard 2D Filters [Reference 11] 168
    2.5.2 Gaussian 2D Filters and Properties 172
    2.5.3 Creating Symmetrical 2D Filters from 1D FIR Filters 178
  Chapter 6: Implementation, Tips and Tricks 187
    2.6.1 Nomenclature 187
    2.6.2 Software Implementation 188
    2.6.3 Fixed Point Implementation 191

    2.6.4 IIR versus FIR 196
    2.6.5 Parallel versus Cascade 197
    2.6.6 Exploiting Symmetry – FIR 200
    2.6.7 Polyphase Structures – FIR 201
    2.6.8 Scaling 202
    2.6.9 Filter Reuse 206
    2.6.10 Real Time Implementations via Our Simple Formulas. 207

Part 3: Advanced Topics 209
  Chapter 7: Linear Phase IIR Filter Design 209
  Chapter 8: Mapping Filters to Different Frequencies 215
  Chapter 9: References 217
  Index 221

This book is for you

I wrote this book to make digital filters more accessible. Practicing engineers will appreciate its straightforward approach and the simple formulas that readily lend themselves to real-time applications. Others will find that digital filter design and analysis is really not as difficult as they may have thought. For each IIR filter type, the reader will find one equation for each coefficient. Plug in what you know – cutoff frequency, sample rate – and the equations will give you the coefficient values; no transforms or complicated manipulations are needed. This approach does have its limitations. Although the book does explain how to create higher orders by combining lower orders, there are no equations for IIR filters larger than fourth order. Several FIR methods are included and they do apply to any order. Also, since elliptical (Cauer) IIR filters and the Remez and Parks-McClellan algorithms for equiripple FIR design require specialized software and do not lend themselves to simple formulas, they are not included herein.

Third Edition

Those of you who have seen previous editions will find this one similar in format to the second, and containing all of the material found in the first and second editions. What I've added are a chapter on two-dimensional filters, and a section on implementing filters in software. Otherwise the book's been edited and reformatted for greater clarity. Hope you have fun!

I did my best but...

I derived most of the equations in the **Handbook** section from scratch, specifically for use in this book. I tested them before including them, then extracted them from the document and tested them again. Through this and other processes, I have done my best to assure there are no errors anywhere in the book; but there surely are. If you find any, let me know and I'll fix them for future printings and editions. In the meantime, if you use these equations to design a filter for your guidance system only to have your whirligig veer off course knocking your grandmother's antique teapot, complete with antique tea, over onto your Sunday paper precipitating inky tea drips upon your freshly tatted doily, you're on your own buddy. In other words, the reader assumes all risk and any damages for use of the materials herein. There is no guarantee, representation, or warranty, real or implied (including merchantability), on the part of the author, copyright owner, or publisher. – *Rusty Allred*

Part 1: Introduction & Theory

Operating on numerical data, digital filters do what their physical namesakes do in water systems and coffee makers. That is, their outputs are modified in some beneficial way with respect to their inputs. In many cases, for example, they separate what is not wanted from what is. Often without even realizing it, we have all applied filters. For example, we apply a simple one when we compute an average.

Filters have many beneficial properties and are used in all electronic devices and in data analysis. In the past their design has been mysterious or complicated, but this book aims to change that. While this section does cover basic filtering theory, the **Handbook** section allows the user to design a wide variety of digital filters without knowing extensive theory or mathematics. Even those who are or will become experts in the field will appreciate the straightforward presentation and the references to additional resources.

Chapter 1: Digital Filter Theory

The **Handbook** section facilitates the design of useful filters without intimate knowledge of the theory presented in this chapter, which can be used as reference, or can be skimmed quickly for familiarity before beginning to design filters. Or, if you want, just jump right in. Filter theory is great fun!

1.1.1 Basic System Theory

Many who would approach filter theory will be conversant with the principles of this section. However, since I have written this book so that anyone can learn to design and use filters, let us review system theory briefly, for those who may not be familiar.

A *digital signal* can be thought of as numbers that arrive to the digital system at some *sample rate*, F_s. Note that the symbol F_s denotes *sample frequency*, which is another name for sample rate. This is also sometimes referred to as *sample interval*, which is technically the time between samples, $1/F_s$.

A digital signal is often denoted x, with each individual sample denoted $x(n)$, where n is the sample number. Refer to the example below, a digital signal with four samples at a sample rate of 1/0.001s = 1000 samples per second.

Digital Signal		
n	t	$x(n)$
0	0s	1.03
1	0.001s	-2.18
2	0.002s	9.32
3	0.003s	8.47

Note that the sample rate has practical implications for digital systems and the way they work, but that, in one sense, it really does not matter to digital filters. If the data comes to the filter at one sample per year or one per nanosecond, the filter still does the same thing. And, speaking strictly mathematically, the filter does not even care if the sample rate is uniform, although the system implementing the filter typically does care. In digital filtering, the sample rate really only has meaning in the interpretation of the filter effect, and there it is everything. Therefore, we will typically speak of the sample rate as being inherently important in the filter while knowing that we can actually implement a filter at one rate, and play the output signal back at another, and that only the playback rate really matters.

One example of this is a file filled with numbers that is processed offline, to be played back later. It does not matter if the processing does not happen in real time, as long as the filters applied are correct for the intended playback rate.

Figure 1 is the block diagram of a digital system. Notice the input, x, and the output, y. Although a digital system is not necessarily a digital filter, a digital filter is one example of a digital system.

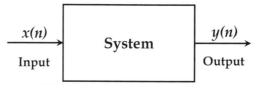

Figure 1 A Digital System

Let us now consider some elements of digital systems. For example, **Figure 2** shows two alternate representations for gain, a common element. The figure also shows the equation: the signal is just multiplied by the gain, as also seen in the example where $a = 2$, as shown at right in the figure.

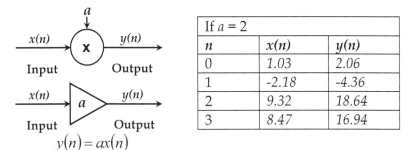

Figure 2 Gain

Another element of a digital system is the delay, as shown in **Figure 3**. Denoted z^{-1}, for reasons which will be clear later, the unit delay element effectively stores the sample for one interval, moving everything back by one sample time. This example assumes the value of the storage element was 0 before the signal began. This is the typical assumption, and is easy to ensure in practice, though it is also possible to do otherwise as needed.

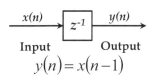

N	x(n)	y(n)
0	1.03	0
1	-2.18	1.03
2	9.32	-2.18
3	8.47	9.32

Figure 3 Delay

Feedback is another important digital system element and can be created using a delay element as seen in **Figure 4**.

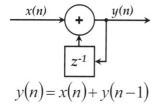

n	x(n)	y(n-1)	y(n)
0	1.03	0	1.03
1	-2.18	1.03	-1.15
2	9.32	-1.15	8.17
3	8.47	8.17	16.64

$y(n) = x(n) + y(n-1)$

Figure 4 Feedback

Now, if we put these elements together, we form a digital system that turns out to be a simple digital filter, as shown in **Figure 5**.

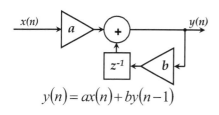

a = 2		b = -0.5	
n	x(n)	y(n-1)	y(n)
0	1.03	0	2.06
1	-2.18	2.06	-5.39
2	9.32	-5.39	21.335
3	8.47	21.335	6.2725

$y(n) = ax(n) + by(n-1)$

Figure 5 Simple Digital Filter

1.1.2 Mathematical Review

As we will quickly encounter below, the mathematics needed for digital filters is primarily high school algebra. Working with filters requires manipulation of polynomials, such as $ax^2 + bx + c$. As discussed more fully below, we commonly need to factor, or find the roots of, these polynomials to understand more about our filters. For a second order polynomial, such as this one, we can find the roots as shown in **Equation 1**.

$$\frac{-b \pm \sqrt{b^2 - 4ac}}{2a} \quad \text{Equation 1}$$

The Quadratic Equation

For example, if $a = 1$, $b = 0.7$, and $c = 0.12$ ($x^2 + 0.7x + 0.12$), the roots must be

$$\frac{-0.7 \pm \sqrt{0.49 - 0.48}}{2} = \frac{-0.7 \pm \sqrt{0.01}}{2} = \frac{-0.7 \pm 0.1}{2} = \frac{-0.7 + 0.1}{2}, \frac{-0.7 - 0.1}{2} = -0.3, -0.4$$

We can then rewrite the polynomial in factored form as follows:

$1x^2 + 0.7x + 0.12 = (x+0.3)(x+0.4)$. Note the form $(x - \text{root}_1)(x - \text{root}_2)$. Of course, it can be a bit difficult for higher-order polynomials; though they can sometimes be factored simply by hand, a good scientific calculator or software such as Matlab or Maple might be required. There are also factoring tools available online.

Another mathematical concept needed in filter work is that of complex numbers, such as $a+bi$. This is because, typically, the roots of polynomials will be complex. Furthermore, we know from mathematics that if a polynomial with real coefficients has complex roots they must occur in complex conjugate pairs. That is, if $a+bi$ is one of the roots, $a-bi$ must be another. How does this give us real coefficients? Watch what happens when we multiply them out: $(x-a-bi)(x-a+bi) = x^2 - 2ax + a^2 + b^2$; the coefficients are real! You will recall that $i^2 = -1$. This principle is used in the product above. These properties of polynomials and complex numbers are important since they apply generally to filters with real coefficients, the subject of this book.

In our filter work we also use complex numbers expressed in exponential notation. This actually sounds more difficult than it turns out to be: $a+bi \leftrightarrow \sqrt{a^2+b^2} \exp\left(i \tan^{-1}(b/a)\right)$.

In addition to these concepts from algebra, a few definitions from trigonometry are also needed for filter work. No need to worry excessively about this since we really do not need trigonometric identities or have to do many manipulations. Rather, we just need to be able to evaluate expressions such as $\cos(x)$, or $\sinh(y)$. This can be done readily on a $10 scientific calculator, in Excel, or using any number of different mathematics software tools.

Throughout this book we will show additional mathematical concepts as needed. If you have an understanding of this section, you will have no problem with the remaining concepts. If not, read it again, Google on it, or contact me through my website: fltrs.com.

1.1.3 Definitions

Before moving on to theory, let us consider some filter definitions. Filters are typically described in terms of their effects as a function of frequency, as shown in **Figure 6** below. A *lowpass* filter, for instance, passes low frequencies and attenuates higher ones. Of course there are other types too, but most can be characterized as one of these types. In these diagrams, the areas of higher magnitudes are frequencies where the signal passes (*passbands*), and those where they are low are frequencies where the signal is decreased (*stopbands*).

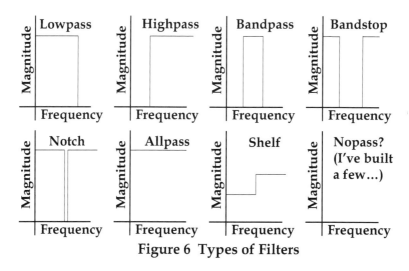

Figure 6 Types of Filters

Actually though, these ideal filters with their abrupt transitions are not realizable. An example of something closer to reality is shown in **Figure 7**. Note the *transition band* and the various definitions.

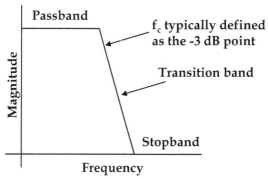

Figure 7 Realizable Lowpass Filter

The *critical frequency* is denoted f_c, and may also be called *center frequency* or *corner frequency* depending upon filter type. In addition, the widths of bandpass, bandstop and notch filters are defined by what is called the bandwidth, *BW*. Some authors also refer to the *Q* of these filters, where $Q = f_c/BW$.

1.1.4 Basic Filter Theory

Unlike their analog counterparts that manipulate voltages and currents, digital systems manipulate numbers, as we discussed previously. Therefore, with a bit of math background, the inner workings of a digital filter are easy to understand.

For example, imagine that you have an Excel spreadsheet, Matlab, or a Big Chief tablet and a number two pencil, and that you would like to compute the average time you spend jogging each day. Perhaps you have noted your total workout times in minutes for the past week and arrived at the following:

6 Days ago	25.34
5 Days ago	25.92
4 Days ago	26.01
3 Days ago	26.23
2 Days ago	26.98
Yesterday	27.20
Today	28.11

Your average workout time is
$$\frac{25.34 + 25.92 + 26.01 + 26.23 + 26.98 + 27.20 + 28.11}{7} = 26.54 \text{ right?}$$

We have all done this many times before, perhaps not the jogging, but at least some sort of averaging. But you might not have realized that the average is a type of digital filter, as further described below.

In the case of our example, the filter input is $x(n)$ = 25.34, 25.92, 26.01, 26.23, 26.98, 27.20, 28.11. We can write the following equation for our averaging filter. Note that $x(n-1)$, $x(n-2)$... indicate one sample ago, two samples ago, and so on:

$$y(n) = \tfrac{1}{7}\sum_{i=0}^{6} x(n-i) = \tfrac{1}{7}x(n) + \tfrac{1}{7}x(n-1) + \tfrac{1}{7}x(n-2) + ... + \tfrac{1}{7}x(n-6)$$

Equation 2

The block diagram of this filter is shown in **Figure 8**.

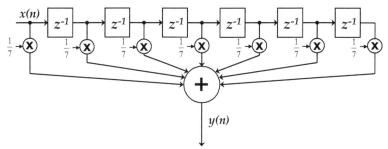

Figure 8 Seven Sample Moving Average Filter

Now let us interpret this filter structure: assume that the delay blocks all begin with stored values of zero, and that our first sample, 25.34, taken 6 days ago, is present on the input. With all the other nodes at 0, the output is 25.34 / 7 = 3.62. This intermediate value, however, is not the value of interest, since it takes into account only our first data sample.

Therefore, we will continue to run our filter every day until it takes into account all seven of our input samples. On the second day, for instance, our first sample has moved into the first storage element making its value present on the second node; the second sample is now on the first node. At this instant we now find that the output of our filter is (25.34 + 25.92) / 7 = 7.32. Again, this is an intermediate value that takes into account only our first two samples. Therefore, we must run the filter for several more samples before we achieve the desired result. Please refer to **Table 1**.

From the table, note the following: a) everything was zeroed before we began, b) sample time n = 6 is the one instant where all seven of our input values are taken into account, and is, therefore, the same instant where the average of our inputs appears at the output, c) if we assume the input goes back to zero after all of our time samples have been used, the value at the output begins to drop off accordingly. If instead we were to continue to feed in valid samples, the output would continue to be a valid average of the last seven samples. This is known as a *moving average filter*.

Note that the filter in **Figure 8** manipulates only the filter inputs to produce an output. This is a characteristic of a moving average filter, which is also known as a *finite impulse response* (FIR) filter. The reason for this name will be discussed in more detail in the **Impulse and Step Responses** section below.

n	x(n)	x(n-1)	x(n-2)	x(n-3)	x(n-4)	x(n-5)	x(n-6)	y(n)
-1	0	0	0	0	0	0	0	0
0	25.34	0	0	0	0	0	0	3.62
1	25.92	25.34	0	0	0	0	0	7.32
2	26.01	25.92	25.34	0	0	0	0	11.04
3	26.23	26.01	25.92	25.34	0	0	0	14.79
4	26.98	26.23	26.01	25.92	25.34	0	0	18.64
5	27.2	26.98	26.23	26.01	25.92	25.34	0	22.53
6	28.11	27.2	26.98	26.23	26.01	25.92	25.34	**26.54**
7	0	28.11	27.2	26.98	26.23	26.01	25.92	22.92
8	0	0	28.11	27.2	26.98	26.23	26.01	19.22
9	...							...

Table 1 Outputs of The Averaging Filter

Now consider the simple filter of **Figure 9** below. Notice, in this case, how the input is manipulated only very little by the filter, but that there is a feedback loop on the output. Since the output of this filter will be fed back and appear in scaled form again at the output in the next time step only to be then fed back again in the next step and the next step and so on, this filter is an *infinite impulse response* (IIR) filter.

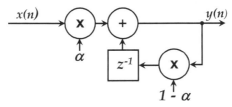

Figure 9 An Infinite Impulse Response (IIR) Filter

Different filters manipulate the inputs and outputs differently, but the primary thing to note is this: if there is feedback from the output, no matter how the input is manipulated, the filter is IIR. Otherwise it is FIR. It is that simple!

1.1.5 The Transfer Function

Comparing **Equation 2** with **Figure 8**, it is easy to see why the output is simply the average of the current input, and the 6 most recent values of that input. Now consider once again **Figure 9**. Here the filter equation is as follows:

$$y(n) = \alpha x(n) + (1-\alpha)y(n-1) \qquad \text{Equation 3}$$

As seen in the figure the unit delay block delays the value of the output by one sample and has no direct effect on the input.

Besides difference equations, such as **Equation 2** and **Equation 3**, there are other mathematical forms used to further understand filter behavior. Chief among these is the *transfer function*, which is a mathematical relationship between the input and the output of a filter. **Figure 10** describes this concept.

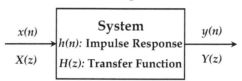

Figure 10 System Response

Relating the system input and output is something called the *impulse response*, which we discuss in more detail below. But, for much of filter analysis, it is mathematically convenient to consider instead the transfer function, which is written in z-transform notation. A deep understanding of z-transform theory is not required; for most filter design work the following rule is sufficient:

$$g(n-\Delta) \rightarrow G(z)z^{-\Delta} \qquad \text{Equation 4}$$

Changing From Difference Equations to z-Transform Notation

That is, starting with the difference equation, change the lower case letters to upper case, and change delays of Δ to multiplicative factors of $z^{-\Delta}$. Let us try it out on our two filter equations above:

$$y(n) = \tfrac{1}{7}\sum_{i=0}^{6} x(n-i) = \tfrac{1}{7}x(n) + \tfrac{1}{7}x(n-1) + \tfrac{1}{7}x(n-2) + \ldots + \tfrac{1}{7}x(n-6)$$

$$Y(z) = \tfrac{1}{7}\sum_{i=0}^{6} X(z)z^{-i} = \tfrac{1}{7}X(z) + \tfrac{1}{7}X(z)z^{-1} + \tfrac{1}{7}X(z)z^{-2} + \ldots + \tfrac{1}{7}X(z)z^{-6}$$

We will focus on the expanded form (without the sum) since it makes the implementation clearer:

$$Y(z) = \tfrac{1}{7}X(z) + \tfrac{1}{7}X(z)z^{-1} + \tfrac{1}{7}X(z)z^{-2} + \ldots + \tfrac{1}{7}X(z)z^{-6} \qquad \text{Equation 5}$$

Averaging Filter (Figure 8) Described in z-Transform Notation

Then, for **Equation 3**: $y(n) = \alpha x(n) + (1-\alpha)y(n-1)$

$$Y(z) = \alpha X(z) + (1-\alpha)Y(z)z^{-1} \qquad \text{Equation 6}$$
Filter of Figure 9 Described in z-Transform Notation

So far pretty straightforward, right? And you can keep smiling because filter theory really never gets much more difficult than this.

Now that we have put our filter equations in z-transform notation, we have added flexibility for manipulation and understanding of these equations. First, let us consider the idea of the transfer function as defined in **Equation 7**:

$$H(z) = \frac{Y(z)}{X(z)} \qquad \text{Equation 7}$$
Definition of Transfer Function

The transfer function, $H(z)$, is that equation by which one can multiply the (z-transform of the) input to get the (z-transform of the) output. Although we are working in z-transform space, as promised above, filtering does not actually require us to take transforms of our inputs and outputs. Rather, we take only the z-transform of the impulse response to form the transfer function, and analyze that to learn much about the filter. Implementation is accomplished entirely in the sample domain.

So what are the transfer functions of our two filter equations above?

$$Y(z) = \tfrac{1}{7}X(z) + \tfrac{1}{7}X(z)z^{-1} + \tfrac{1}{7}X(z)z^{-2} + \ldots + \tfrac{1}{7}X(z)z^{-6}$$
$$Y(z) = \tfrac{1}{7}X(z)\left[1 + z^{-1} + z^{-2} + z^{-3} + z^{-4} + z^{-5} + z^{-6}\right]$$

$$H(z) = \frac{Y(z)}{X(z)} = \tfrac{1}{7}\left[1 + z^{-1} + z^{-2} + z^{-3} + z^{-4} + z^{-5} + z^{-6}\right] \qquad \text{Equation 8}$$
Transfer Function of the Averaging Filter of Figure 8

$$Y(z) = \alpha X(z) + (1-\alpha)Y(z)z^{-1}$$
And for **Equation 6**: $\quad Y(z)\left[1 - (1-\alpha)z^{-1}\right] = \alpha X(z)$
$$Y(z)\left[1 + (\alpha-1)z^{-1}\right] = \alpha X(z)$$

$$H(z) = \frac{Y(z)}{X(z)} = \frac{\alpha}{1+(\alpha-1)z^{-1}}$$ Equation 9

Transfer Function of the IIR Filter of Figure 9

What is remarkable when comparing these two transfer functions? Note that **Equation 9** is a rational function; that is its denominator is a function of z^{-1}. Recall that **Equation 9** relates to **Figure 9**, an IIR filter, and that **Equation 8** relates to **Figure 8**, an FIR filter. The difference we see in these two equations is general: an IIR filter will always have a denominator that is a function of z^{-1}. (The numerator will generally be a function of z^{-1} too, but does not have to be.) An FIR filter will typically have a denominator of 1 (seldom shown) and can never have a denominator that is a function of z^{-1}.

Before moving on, let us try our new tools on one more filter, as shown in **Figure 11**. First, by observation, what kind of filter is this? Second, what is its difference equation description? Third, what is its transfer function?

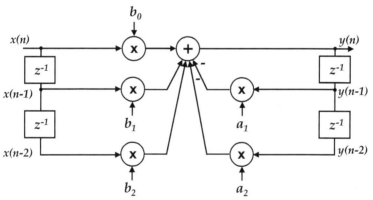

Figure 11 New Mystery Filter

As discussed previously, the filter of **Figure 11** must be IIR since it has feedback from the output. Looking backward from its output, it is easy to construct the difference equation, **Equation 10**:

$$y(n) = b_0 x(n) + b_1 x(n-1) + b_2 x(n-2) - a_1 y(n-1) - a_2 y(n-2)$$ Equation 10

Now, using **Equation 4**, convert this to z-transform notation:

$$Y(z) = b_0 X(z) + b_1 X(z) z^{-1} + b_2 X(z) z^{-2} - a_1 Y(z) z^{-1} - a_2 Y(z) z^{-2}$$

$$Y(z)\left[1 + a_1 z^{-1} + a_2 z^{-2}\right] = X(z)\left[b_0 + b_1 z^{-1} + b_2 z^{-2}\right]$$

$$\frac{Y(z)}{X(z)} = \frac{b_0 + b_1 z^{-1} + b_2 z^{-2}}{1 + a_1 z^{-1} + a_2 z^{-2}}$$

$$H(z) = \frac{b_0 + b_1 z^{-1} + b_2 z^{-2}}{1 + a_1 z^{-1} + a_2 z^{-2}} \qquad \text{Equation 11}$$

Transfer Function of the 2nd-order IIR Filter of Figure 11

As we knew already due to the feedback, this is an IIR filter. And, since its maximum delay is 2, we call it a second-order IIR filter.

1.1.6 Filter Structures

Figure 8 above is a type of simple FIR filter. While they are sometimes drawn differently, any filter that manipulates only the input signal to form the output will be FIR. A general N^{th}-order FIR filter is shown in **Figure 12**. Note that an Nth-order filter has $N+1$ coefficients (b_n) because one is applied to the current (delay free) input, and then one to each of the delayed input values.

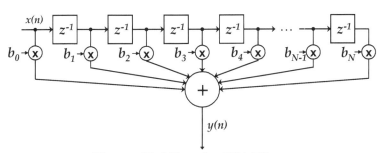

Figure 12 N^{th}-order FIR Filter

In general, the coefficients are constrained only in that they must be producible by the implementation architecture, though linear phase FIR filters, which are discussed in detail in a later chapter, have additional constraints. The difference equation for the general FIR is:

$$y(n) = \sum_{i=0}^{N} b_i x(n-i) = b_0 x(n) + b_1 x(n-1) + b_2 x(n-2) + \ldots + b_N x(n-N) \qquad \text{Equation 12}$$

General FIR Difference Equation

And the transfer function is:

$$H(z) = b_0 + b_1 z^{-1} + b_2 z^{-2} + b_3 z^{-3} + \cdots + b_N z^{-N}$$ **Equation 13**

General FIR Transfer Function

IIR filters have a wide variety of forms. The second-order filter of **Figure 11** is in what is known as Direct Form I. Such a filter can be of any order, as indicated by **Figure 13**.

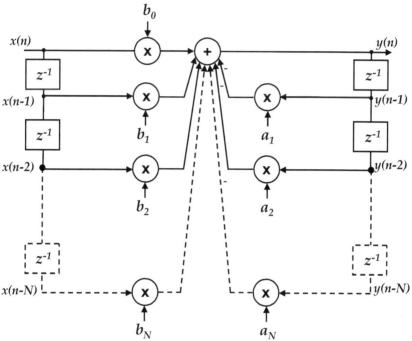

Figure 13 N^{th}-order Direct Form 1 IIR Filter

The transfer function for this filter is given by **Equation 14**:

$$H(z) = \frac{b_0 + b_1 z^{-1} + b_2 z^{-2} + \cdots + b_N z^{-N}}{1 + a_1 z^{-1} + a_2 z^{-2} + \cdots + a_N z^{-N}}$$ **Equation 14**

General IIR Transfer Function

There are other implementation structures which share this same transfer function, including two variants of Direct Form II shown in **Figure 14** and **Figure 15**.

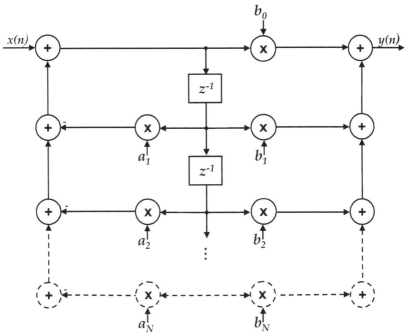

Figure 14 N^{th}-order Direct Form 2 IIR Filter

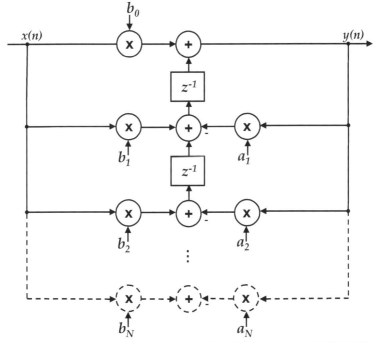

Figure 15 Alternate Form N^{th}-order Direct Form 2 IIR Filter

Another type of filter that is encountered occasionally is the lattice, as shown in **Figure 16**. This filter has the difference equation and transfer function shown in **Equation 15** and **Equation 16**.

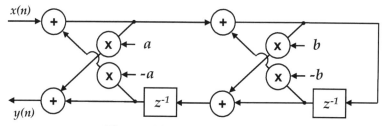

Figure 16 Lattice IIR Filter

$$y(n) = ax(n) + b(a+1)x(n-1) + x(n-2) - b(a+1)y(n-1) - ay(n-2)$$

Equation 15

$$H(z) = \frac{a + b(a+1)z^{-1} + z^{-2}}{1 + b(a+1)z^{-1} + az^{-2}}$$

Equation 16

There are a couple of unusual things to notice about this filter: first the feedback from y(n) is not obvious at first, but it does exist. Notice how there are feedback paths between the top and bottom rails of the filter. In fact, the input into the summer just right of y(n) is all of y(n) except for a scaled copy of x(n). That is, a signal that is very nearly y(n) is being fed back and summed with x(n) at that point. In the end, the filter does have feedback and is IIR. Analyzing the difference equation for this filter is not trivial. For more information on that process, please refer to **Reference 5** pages 70 – 78.

In addition, notice the similarity between the numerator coefficients and the denominator coefficients. When an IIR filter's numerator has coefficients that are a mirror image of its denominator coefficients, that filter is called *allpass*. Allpass filters are useful for changing phase without changing magnitude. To turn this filter into something that can also change magnitude, we can modify the structure as shown in **Figure 17**. (Allpass filters, phase responses, and magnitude responses are all discussed in more detail in subsequent sections.)

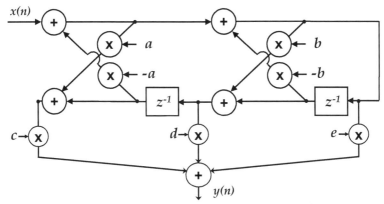

Figure 17 Lattice-based General IIR Filter

Notice that the output has been moved to a different point, and that additional manipulations are carried out on intermediate points as well. The difference equation and transfer function for this filter are shown in **Equation 17** and **Equation 18**.

$$y(n) = (ac + bd + e)x(n) + (abc + bc + d)x(n-1) + cx(n-2) - b(a+1)y(n-1) - ay(n-2)$$

Equation 17

$$H(z) = \frac{ac + bd + e + (abc + bc + d)z^{-1} + cz^{-2}}{1 + b(a+1)z^{-1} + az^{-2}}$$

Equation 18

Notice that the scale factors on *x* have changed in the difference equation and the numerator of the transfer function has changed accordingly. Now the numerator and denominator coefficients are no longer symmetrical and the filter is no longer an allpass filter (unless *c* is 1 and *d* and *e* are both 0).

1.1.7 Filter Analysis

In previous sections we mentioned that z-transform notation is used to characterize filters. In this section we discuss how it is beneficial in understanding filter characteristics.

1.1.7.1 Poles, Zeros and Stability

The heart of filter analysis is an understanding of the roots of the polynomials making up the filter transfer functions. These roots give us important information about the associated filters. Let us consider our moving average filter with this transfer function:

$$H(z) = \tfrac{1}{7}\left[1 + z^{-1} + z^{-2} + z^{-3} + z^{-4} + z^{-5} + z^{-6}\right]$$

Before finding the roots of filter polynomials, we always put the polynomial in terms of z, rather than z⁻¹. This is *analysis form*, whereas the version in terms of z⁻¹ is *implementation form*. Note that so long as we multiply the numerator and the denominator by the same value, the ratio remains the same. So let us first make our transfer function explicitly a ratio by showing the denominator of 1:

$$H(z) = \frac{\tfrac{1}{7}\left[1 + z^{-1} + z^{-2} + z^{-3} + z^{-4} + z^{-5} + z^{-6}\right]}{1}$$

Now, we can get this in terms of z by multiplying by $\dfrac{z^6}{z^6}$; since we are effectively multiplying by 1, the equality holds.

$$H(z) = \frac{\tfrac{1}{7}\left[1 + z^{-1} + z^{-2} + z^{-3} + z^{-4} + z^{-5} + z^{-6}\right]}{1} \cdot \frac{z^6}{z^6}$$

$$H(z) = \frac{\tfrac{1}{7}\left[z^6 + z^5 + z^4 + z^3 + z^2 + z^1 + 1\right]}{z^6}$$

With me so far? All we have done is put this in standard analysis notation without changing the value of the ratio. Note that the function of z in the denominator does not mean this is now an IIR filter. That only applies when the transfer function is written in terms of z⁻¹, which is the standard implementation notation (since it explicitly shows the filter delays), while putting things in terms of z is essential for understanding the roots of the polynomials.

Now that we are in standard analysis notation, let us introduce the terminology *zeros* for the roots of numerator polynomials, and *poles*, for the roots of denominator polynomials. What can we say immediately about the poles of this filter? $z^6 = z \cdot z \cdot z \cdot z \cdot z \cdot z$ right? Or it could be written $(z-0)(z-0)(z-0)(z-0)(z-0)(z-0)$. In other words, the poles are all real, and are all equal to 0. This is true for all FIR filters. Now let us consider the zeros. Using a root finder we

discover that the roots of the numerator polynomial are:
 0.623489801858734 ± 0.781831482468030i
 -0.900968867902419 ± 0.433883739117558i
 -0.222520933956315 ± 0.974927912181823i

These are conjugate pairs, as discussed earlier. Now that we have computed the poles and zeros, we are ready to plot them in the *complex plane*, and also to discuss the idea of stability.

Figure 18 plots our poles and zeros in what we call the *z-plane*, which is just a complex plane where the x-axis is real and the y-axis is imaginary. Note that our poles are indicated by x on the plot and that there are 6 of them all at zero, as discussed above. The zeros are indicated by o and are distributed around the circle. Note that complex conjugate pairs are mirror images about the real axis.

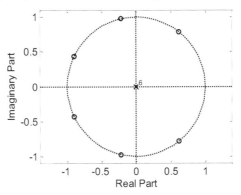

Figure 18 Poles and Zeros of Our 6th-order Moving Average Filter

The large circle in the plot is called the unit circle. That is, it has a radius of 1 in this plane. This is important due to insight we gain from knowing how the poles lie with respect to this circle. *For a stable filter, the poles must lie strictly inside the unit circle.* Poles directly on the circle form an oscillator (output varies in a regular pattern); this is marginal stability. Poles outside the circle indicate strict instability. Note that zeros do not affect stability and can be inside, on, or outside the circle for a stable filter.

We will explore these concepts further using our simple filter from **Figure 9** above. This filter is sometimes called an alpha filter, due to its coefficient α. Sometimes it is also called an averaging filter, though it behaves rather differently than the moving average filter of our previous examples. Either way, the function of the alpha filter is to form a balance between the current input sample and the collected filter history y(n-1). α must have a value between 0 and 1. If it is 0, the input cannot affect the output at all. If it is 1, the input passes through without modification. Anything in between is some blend between the current input sample and the previous output sample. Let us use our newfound filter analysis abilities on this filter.

First, recall that the transfer function (**Equation 9**) is:

$$H(z) = \frac{\alpha}{1+(\alpha-1)z^{-1}}$$

To put it in analysis form we simply multiply numerator and denominator each by z:

$$H(z) = \frac{\alpha z}{z+(\alpha-1)}$$

As a first order filter we know immediately that its roots must be real since they have no opportunity to occur in complex conjugate pairs. Furthermore, its zero is at 0, and its pole is at 1 - α (think of it as the value of z that causes the denominator to go to 0).

Quickly we see that if α is kept between 0 and 1, as stipulated, the pole will vary between 1 (marginally stable) and 0 (stable). If, however, we were to let α get larger than 1, the filter would be unstable. Let us try a few examples:

Example 1:

α = 0.5, $H(z) = \frac{0.5z}{z-0.5}$. The zero is still at 0, and the pole is at 0.5. Recall that the difference equation for this filter is $y(n) = \alpha x(n) + (1-\alpha)y(n-1)$ (**Equation 3**) generally and, specifically, $y(n) = 0.5x(n) + 0.5y(n-1)$. Now let us see what happens if we receive a step input, assuming y(n-1) is zero; the output is tabled in **Figure 19** next to the pole-zero plot for this case.

Chapter 1 Introduction & Theory

N	x(n)	y(n-1)	y(n)
-1	0	0	0
0	1	0	0.5
1	1	0.5	0.75
2	1	0.75	0.875
3	1	0.875	0.9375

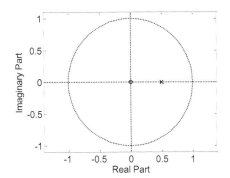

Figure 19 Pole and Zero Plot of Our 1st-order Alpha Filter, α = 0.5

Notice that if we make α = 1, $H(z) = \frac{z}{z}$. This filter's output always equals its input. On the other hand, if α = 0, $H(z) = \frac{0}{z-1}$. Now, the pole indicates marginal stability, but the filter will not respond to any input, since all inputs are multiplied by 0: $y(n) = 0 \cdot x(n) + 1 \cdot y(n-1) = y(n-1)$. Therefore, if this filter starts with an output of 0, it will always have an output of 0. However, if it were started with any other output, that output would remain the output forever, independent of input.

Example 2: α = 2, $H(z) = 2z/(z+1)$, $y(n) = 2x(n) - y(n-1)$. This is a case of marginal stability with the pole moving from +1, for the case of α = 0, to -1, still on the unit circle. Below is the output for this filter, assuming it starts with 0 and has a steady input of 1. This is an oscillator; the output alternates between 0 and 2.

n	x(n)	y(n-1)	y(n)
-1	0	0	0
0	1	0	2
1	1	2	0
2	1	0	2
3	1	2	0

Example 3: α = 3, $H(z) = 3z/(z+2)$, $y(n) = 3x(n) - 2y(n-1)$. Now the pole is distinctly outside the unit circle as seen in **Figure 20**. The table shows that the output grows with each new input sample; it will continue to grow indefinitely, the very definition of instability.

Digital Filters for Everyone

n	x(n)	y(n-1)	y(n)
-1	0	0	0
0	1	0	3
1	1	3	-3
2	1	-3	9
3	1	9	-15
4	1	-15	33
5	1	33	-63
6	1	-63	129
7	1	129	-255

Figure 20 Pole and Zero Plot of Our 1st-order Alpha Filter, α = 3

Now, before going on, let us try an example using a different filter.

Example 4: $H(z) = \dfrac{b_0 + b_1 z^{-1} + b_2 z^{-2}}{1 + a_1 z^{-1} + a_2 z^{-2}}$ (second-order IIR filter).

In particular, let us consider a Butterworth lowpass filter with a cutoff frequency of 2 kHz designed for a sample rate of 10 kHz. We will learn to design this filter in the **Handbook** section. For now we shall stipulate that the coefficients are

b_0 = 0.206572083826148 a_1 = -0.369527377351241
b_1 = 0.413144167652296 a_2 = 0.195815712655833
b_2 = 0.206572083826148

The zeros and poles are plotted in **Figure 21**. The zeros are both -1; the poles are 0.184763688676 ± 0.402092143672i.

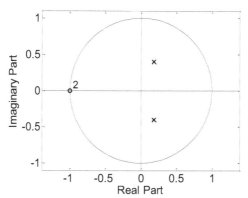

Figure 21 Pole and Zero Plot of Our
2nd-order Butterworth Filter

Chapter 1 Introduction & Theory

Recall that the magnitude of a complex number is given by **Equation 19** which denotes magnitude, r, of the ith pole.

$$r = \sqrt{\text{Re}(P_i)^2 + \text{Im}(P_i)^2}$$ **Equation 19**

Magnitude of a Complex Number

For Example 4:

$$0.442510692137 = \sqrt{0.184763688676^2 + 0.402092143672^2}$$

Of course, the magnitudes of each member of a conjugate pair are always equal.

Figure 22 repeats the pole-zero plot from **Figure 21** but this time with the magnitude and angle information shown. Note that stability only requires that the poles be strictly inside the unit circle, which means that the filter is stable so long as the magnitudes of all poles are less than one: $r_i < 1$.

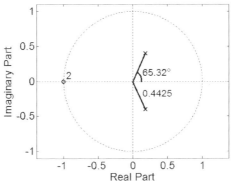

Figure 22 Pole and Zero Plot Showing Polar Coordinates

To understand stability we do not need to know the angle, but to describe the poles using polar coordinates we need that angle along with the magnitude. The angle is computed using **Equation 20** where arctan2 is the four quadrant or two argument arctangent.

$$\theta = \frac{180}{\pi} \arctan 2\left(\text{Im}[P_i], \text{Re}[P_i]\right)$$ **Equation 20**

Phase of a Complex Number

23

One final point about the z-plane: Note that digital filters are defined on a frequency band between 0 Hz and half of the sample rate, or $F_s/2$, and that this range is mapped linearly around the unit circle, between 0 and π. Half the sample rate is called the *Nyquist Frequency* and is the highest possible frequency that a digital filter can affect, under typical sampling situations.

In the example case of **Figure 22** above, note that the angle is 65.32°. Further note that the filter in question has a sample rate of 10 kHz. Since the range of 0 Hz to 5 kHz (half the sample rate) maps to 0 to π (180°), this angle maps to 65.32°/180° x 5000 = 1.814 kHz. This makes sense since the cutoff frequency of the filter is 2 kHz; the poles will not necessarily be right at the cutoff frequency, but we should see activity in that region as is the case here.

1.1.7.2 Filter Characteristics

Using the analysis form of the transfer function, we have been able to determine whether a filter is stable. There are numerous other filter characteristics that we can also determine, as discussed in the sections below.

1.1.7.2.1 Frequency Response and Group Delay

To begin with, let us consider the idea of a frequency response. As shown graphically in **Figure 6** in a previous section, filters are used to create some sort of effect on the output data, and this effect can typically be understood in terms of frequency. The frequency response of a digital filter takes place with respect to the sample rate at which the data passes through the filter or is played back, in the case of offline processing. That is, digital filters create the same interrelationships between output samples independent of sample rate; data passing through the filter at one sample per second will be modified on a sample-by-sample basis in exactly the same way as a signal that passes through at one million samples per second. What will be different, however, is what this playback rate means in terms of actual physical frequency response of the filter on the data.

Due to this property, the frequency characteristics of digital filters are either plotted in a normalized fashion, independent of sample rate, or they are plotted with respect to the intended sample rate. In this book we will primarily use the latter approach since it is the most straightforward for a typical design situation. However, in this section we will also show some normalized plots since the reader will encounter them often in other sources, and since they do make clear the fact that a digital filter's frequency response always depends upon sample rate.

To begin with, let us define the frequency response of a digital filter:

$$H(\omega) = H(z)\big|_{z=e^{i\omega}} \quad \omega = \frac{2\pi f}{F_s} \quad \text{Equation 21}$$

Frequency Response of a Digital Filter

Once again we see from this equation that we can either analyze the frequency response in terms of normalized frequency, ω, or in terms of actual frequency, f. The latter, however, depends upon the value of sample rate, F_s, as we established above.

Let us consider the case of the 2nd-order IIR filter of **Equation 11**:

$$H(z) = \frac{b_0 + b_1 z^{-1} + b_2 z^{-2}}{1 + a_1 z^{-1} + a_2 z^{-2}}$$

$$H(\omega) = \frac{b_0 + b_1 e^{-i\omega} + b_2 e^{-i2\omega}}{1 + a_1 e^{-i\omega} + a_2 e^{-i2\omega}}$$

$$H(f) = \frac{b_0 + b_1 e^{-i\frac{2\pi f}{F_s}} + b_2 e^{-i\frac{4\pi f}{F_s}}}{1 + a_1 e^{-i\frac{2\pi f}{F_s}} + a_2 e^{-i\frac{4\pi f}{F_s}}}$$

For filter analysis there are three useful plots deriving from these equations: the magnitude plot shows how the filter amplifies or attenuates a signal as a function of frequency; the phase plot shows how phase changes as a function of frequency, an often undesirable but inevitable result of filtering; finally, the group delay, which is related to the phase, shows how much delay a signal encounters on its way through a filter, as a function of frequency.

These three entities are defined mathematically by **Equation 22**, **Equation 23**, and **Equation 24**. Note that the absolute value of a complex value is the magnitude, and that arctan2 is the four quadrant or two argument arctangent as was also used above for polar coordinates of poles and zeros.

$$M(\omega) = |H(\omega)|$$
Magnitude Response

Equation 22

$$P(\omega) = \arctan 2(\operatorname{Im}[H(\omega)], \operatorname{Re}[H(\omega)])$$
Phase Response

Equation 23

$$D_G(\omega) = \frac{-dP(\omega)}{d\omega}$$
Group Delay

Equation 24

The easiest way to create these plots is to use a tool that can manipulate complex numbers. In fact, some tools, such as the Signal Processing Toolbox in Matlab, can create all of these plots using the FREQZ command for magnitude and phase, and the GRPDELAY command for group delay. Let us try an example that way, and then discuss options for when a similar tool is unavailable.

As an example, let us try our Butterworth filter from above, which we will recall has the coefficients below:

b_0 = 0.206572083826148 a_1 = -0.369527377351241
b_1 = 0.413144167652296 a_2 = 0.195815712655833
b_2 = 0.206572083826148

Here is what we would do in Matlab:
B = [0.206572083826148 0.413144167652296 0.206572083826148]
A = [1 -0.369527377351241 0.195815712655833]   %(a_0 is 1)
freqz(B, A)

Figure 23 is the default result of the FREQZ command. Notice that it delivers the magnitude plot in dB (more on this below), the phase in degrees, and plots everything on a linear frequency axis. It is also possible to use the FREQZ command to compute the information to be plotted, and then plot it in different ways. For instance, the following command loads the variables H and F with the transfer function, evaluated at various normalized frequencies: [H, F] = freqz(B, A). H and F can then be plotted in any desired fashion. However, we might actually prefer to scale the frequency axis by our sample rate, and tell FREQZ where we would like our frequency samples taken. The following command does that:
[H, F] = freqz(B, A, logspace(0, log10(5000), 350), 10000);

Chapter 1 Introduction & Theory

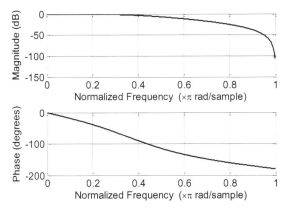

Figure 23 Magnitude and Phase Plots for our Butterworth Filter Example

Recall that we designed this filter for a sample rate of 10 kHz. The command above tells FREQZ to use 350 points on logarithmic spacing between 10^0 (1) and $10^{\log 10(5000)}$ (5000, half the sample rate) and to scale the plot by the sample rate of 10000. Now we can plot this on a logarithmic frequency axis as follows: semilogx(F, 20 * log10(abs(H))). This is a fairly standard way to plot a magnitude response, as seen in **Figure 24**.

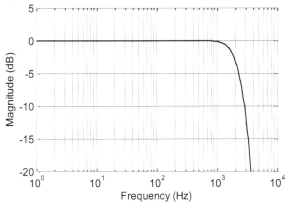

Figure 24 Magnitude Plot for our Butterworth Filter Example

Recall also that we designed it to have a 2 kHz cutoff, and note on the plot that it goes through -3dB at 2 kHz.

If you are not used to the dB scale, it is just a convenient method of showing a large magnitude range on a small scale. Since it is $20\log_{10}$

of the magnitude, if the magnitude goes from 1 to 0.01, as in our plot above, the dB scale will go from 0 to -20. As seen in our Matlab code, we are working with the magnitude of H, since it is complex.

Now let us plot the group delay using the command grpdelay(B, A), the output of which is plotted in **Figure 25**. This plot shows how many samples of delay are encountered by various frequencies as they traverse the filter.

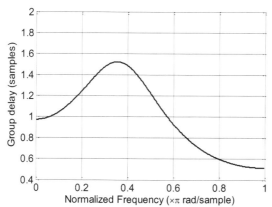

Figure 25 Group Delay for our Butterworth Filter Example

Although there are many powerful filtering tools available, not everyone has access to them, and they are really not required for filter design and analysis. First let us replicate our plots above using Matlab without the Signal Processing Toolbox. This would be similar to using any good mathematics tool.

```
Fs = 10000;
f = 0:4999;
H = (B(1) + B(2) .* exp(-2 .* i .* pi .* f ./ Fs) + B(3) .*
exp(-4.*i.*pi.*f./Fs)) ./ (A(1)+A(2) .* exp(-2.*i.*pi.*f ./ Fs)+A(3) .*
exp(-4.*i.*pi.*f ./ Fs));
semilogx(f, 20*log10(abs(H))) % Produces top plot
plot(f, 180.*atan2(imag(H), real(H)) ./ pi) % Center plot
plot(f(1:end-1)+0.5, -diff(5000 .* atan2(imag(H), real(H)) ./
pi)) %Produces left plot
```

The plots produced by these commands can be found in **Figure 26**. Note that, in addition to the commands used above, Matlab does have an ANGLE command that can be used to produce the phase plot, rather than ATAN2. However, since math tools will generally have an arctan2 command, but may not have an equivalent of ANGLE, we have used ATAN2 for generality. Note also that the phase would be in radians if we had not divided by π and multiplied by 180 to convert to degrees. Finally, note that the DIFF command just takes the difference between two phase points to approximate a derivative.

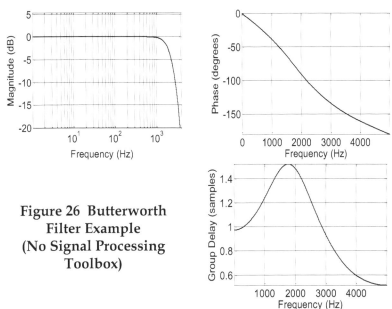

Figure 26 Butterworth Filter Example (No Signal Processing Toolbox)

To scale the group delay in samples, one must divide out π and multiply by the number of points. Note also that since the difference of phase values is one element shorter than the phase vector itself, the frequency vector was also shortened to compensate. The addition of 0.5 to the frequency vector shifts it to the right by a half Hertz to make the frequencies correspond to the midpoint between the two phase values used in the pseudo-differential.

These days there are many tools available for manipulating complex numbers, including excellent calculators. However, one tool that is surprisingly powerful, and to which many people have access, is Microsoft Excel. Now let us create all of these plots using that tool.

To begin, check to be sure the Analysis Tool Pack is enabled under Tools, Add-ins in Excel. This enables a host of engineering functions that are not defaulted on in typical installations. Now we are ready to start creating filter analysis tools. There are many ways to do this; one is shown.

Use row 1 for labels, starting data at row 2. From row 2 down, make your frequency vector, from 0 to 4999. Next to that make the vector that will be used to plot the group delay by adding 0.5 to the frequency vector and going to one cell before the bottom. These columns will appear as shown below. Off to the right in the worksheet, make cells for inputting the filter coefficients and the sample rate, as shown:

A	B	...	J	K	L
F	fshift	...	B	A	Fs
0	0.5	...	0.206572084	1	10000
1	1.5	...	0.413144168	-0.369527377	
2	2.5	...	0.206572084	0.195815713	
3	3.5	...			
4	4.5	...			
5	5.5	...			
6	6.5	...			
...	...	...			
4998	4998.5	...			
4999		...			

Now, since they will be used numerous times in each computation, it is a good idea to set up columns for $e^{-i\frac{2\pi f}{F_s}}$ and $e^{-i\frac{4\pi f}{F_s}}$:

Column C: = IMEXP(IMPRODUCT("1i", -2*PI()*A2/$L$2)) and
Column D: = IMEXP(IMPRODUCT("1i", -4*PI()*A2/$L$2))

Note that Excel has special functions for manipulating complex numbers: IMEXP is the exponential, IMPRODUCT is the multiply, IMSUM is the sum and IMDIV is the divide. Also note how the complex variable i is input, and that $L$2 is applied to input F_s so that it stays constant as the formula is replicated.

Now it is time to create the transfer function, H in Column E:
= IMDIV(IMSUM($J$2, IMPRODUCT($J$3, C2), IMPRODUCT($J$4, D2)),
IMSUM($K$2, IMPRODUCT($K$3, C2), IMPRODUCT($K$4, D2)))

Again the dollar sign is used to keep the coefficients fixed as the formula is replicated.

Chapter 1 Introduction & Theory

C	D	E
1st	2nd	H
1	1	1
0.9999998-6.2831848E-004i	0.9999992-1.2566367E-003i	0.9999998-6.1151040E-004i
0.9999992-1.2566367E-003i	0.9999968-2.5132715E-003i	0.9999993-1.2230209E-003i
0.9999982-1.8849545E-003i	0.9999929-3.7699023E-003i	0.9999983-1.8345317E-003i
0.9999968-2.5132715E-003i	0.9999874-5.0265271E-003i	0.9999970-2.4460428E-003i
...	...	...
-0.9999992-1.2566367E-003i	0.9999968+2.5132715E-003i	-2.0839238E-007-1.3453452E-010i
-0.9999998-6.2831849E-004i	0.9999992+1.2566367E-003i	-5.2098094E-008-1.6815681E-011i

From this column we can create columns for the 3 different plots:

Column F: =20*LOG10(IMABS(E2))
Column G: =180*ATAN2(IMREAL(E2), IMAGINARY(E2))/PI()
Column H: =5000*(G2-G3)/180

F	G	H
20log10(|H|)	Phase	Group Delay
0	0	0
-1.46578E-13	-0.035036967	0.973249638
-2.43107E-12	-0.070073954	0.97325075
-1.22981E-11	-0.105110981	0.973252418
-3.8872E-11	-0.140148068	0.973254643
...	...	...
-133.6223614	-179.9630108	0.513772897
-145.6635629	-179.9815067	

Since the phase column was scaled in degrees ($180/\pi$), the group delay, was scaled by 5000/180, rather than by 5000. **Figure 27** shows the plots.

Note that the magnitude plot is on a linear frequency scale, so it will compare best to **Figure 23** above, although that plot is in normalized frequency units.

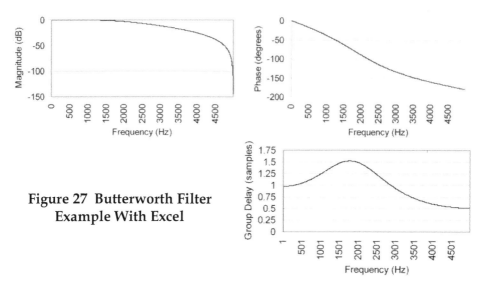

Figure 27 Butterworth Filter Example With Excel

In summary, while there are many powerful filtering tools available, any tool that can manipulate complex values and create graphs is adequate for analyzing filters.

1.1.7.2.2 Impulse and Step Responses and Transient Analysis

Besides the frequency domain filter analyses detailed above, there are also time domain analyses that turn out to be very useful. Chief among these are the impulse response and the step response.

As seen in **Figure 28**, a digital impulse signal has a value of 1 at some point (usually sample 0), and is 0 at every other sample time.

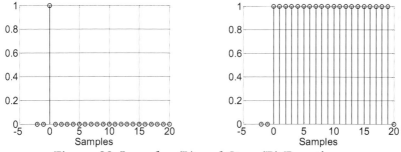

Figure 28 Impulse (L) and Step (R) Functions

Similarly, the step signal is valued 0 at all previous samples, transitions to 1 at some point (usually sample 0), and is then 1 for all samples to the right. These two signals have particular value in understanding filter behavior.

The impulse response and step response are merely a filter's responses to these two inputs. Let us try some examples; first of our averaging filter in **Figure 8**.

Recall that its difference equation is

$$y(n) = \tfrac{1}{7}x(n) + \tfrac{1}{7}x(n-1) + \tfrac{1}{7}x(n-2) + \ldots + \tfrac{1}{7}x(n-6)$$

From this equation, and as also seen in the figure, as the impulse shifts through the delays, the value of each coefficient appears in turn at the output. The impulse response is plotted in **Figure 29 (L)**.

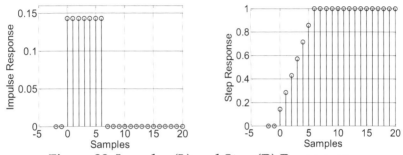

Figure 29 Impulse (L) and Step (R) Responses

Notice what happens as the step function moves through the averaging filter. Recall that the step will remain at value 1 forever, once it begins. Therefore, as the first sample of the step moves into the filter, the output will be b_0, or 1/7 in this specific case. The second sample results in $b_0 + b_1$ or 1/7 + 1/7 = 2/7 and so forth. When the first 7 samples have shifted into our averaging filter, the output will be 1, and it will remain 1 forever as a new input of 1 shifts into the input as each old sample of 1 shifts out. The result is shown graphically in **Figure 29 (R)**.

As is clear from studying the difference equation and the filter structure, it is a characteristic of FIR filters that the impulse response will always be a replica of the filter coefficients. And, as we have already established, this impulse response is finite in duration, just as there is a finite number of coefficients.

Regarding the step response, this indicates transient behavior of the filter. As a signal hits the filter, it takes it this long to "charge," as we have already established.

The Signal Processing Toolbox in Matlab has IMPZ and STEPZ commands for simply computing impulse and step responses from coefficients. However, it is also a simple matter to derive these from the difference equations as we have done in these examples.

Now, let us consider the case of the alpha filter of **Figure 9**, which is described by the following difference equation:

$$y(n) = \alpha x(n) + (1-\alpha)y(n-1).$$

What are the impulse and step responses for this filter for values of $\alpha = 0.25$ and $\alpha = 0.1$? For the first case, the difference equation becomes: $y(n) = 0.25x(n) + 0.75y(n-1)$. The first twenty samples of the impulse and step responses are tabled below.

Impulse Response ($\alpha = 0.25$)				Step Response		
x(n)	y(n-1)	y(n)		x(n)	y(n-1)	y(n)
1	0	0.25		1	0	0.25
0	0.25	0.1875		1	0.25	0.4375
0	0.1875	0.140625		1	0.4375	0.578125
0	0.140625	0.105469		1	0.578125	0.683594
0	0.105469	0.079102		1	0.683594	0.762695
0	0.079102	0.059326		1	0.762695	0.822021
0	0.059326	0.044495		1	0.822021	0.866516
0	0.044495	0.033371		1	0.866516	0.899887
0	0.033371	0.025028		1	0.899887	0.924915
0	0.025028	0.018771		1	0.924915	0.943686
0	0.018771	0.014078		1	0.943686	0.957765
0	0.014078	0.010559		1	0.957765	0.968324
0	0.010559	0.007919		1	0.968324	0.976243
0	0.007919	0.005939		1	0.976243	0.982182
0	0.005939	0.004454		1	0.982182	0.986637
0	0.004454	0.003341		1	0.986637	0.989977
0	0.003341	0.002506		1	0.989977	0.992483
0	0.002506	0.001879		1	0.992483	0.994362
0	0.001879	0.001409		1	0.994362	0.995772
0	0.001409	0.001057		1	0.995772	0.996829
0	0.001057	0.000793		1	0.996829	0.997622

As established earlier, this is an IIR filter, so the impulse response is technically infinite. As you can see from the table, the impulse response has not gone identically to zero, though the bulk of the energy is in the first dozen samples or so. The responses are plotted in **Figure 30**.

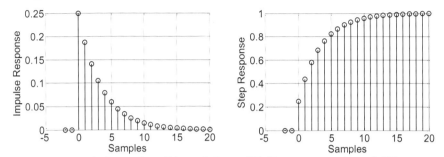

Figure 30 Impulse (L) and Step (R) Responses ($\alpha = 0.25$)

For the second case, $y(n) = 0.1x(n) + 0.9y(n-1)$.

Impulse Response ($\alpha = 0.1$)			Step Response		
x(n)	y(n-1)	y(n)	x(n)	y(n-1)	y(n)
1	0	0.1	1	0	0.1
0	0.1	0.09	1	0.1	0.19
0	0.09	0.081	1	0.19	0.271
0	0.081	0.0729	1	0.271	0.3439
0	0.0729	0.06561	1	0.3439	0.40951
0	0.06561	0.059049	1	0.40951	0.468559
0	0.059049	0.053144	1	0.468559	0.521703
0	0.053144	0.04783	1	0.521703	0.569533
0	0.04783	0.043047	1	0.569533	0.61258
0	0.043047	0.038742	1	0.61258	0.651322
0	0.038742	0.034868	1	0.651322	0.686189
0	0.034868	0.031381	1	0.686189	0.71757
0	0.031381	0.028243	1	0.71757	0.745813
0	0.028243	0.025419	1	0.745813	0.771232
0	0.025419	0.022877	1	0.771232	0.794109
0	0.022877	0.020589	1	0.794109	0.814698
0	0.020589	0.01853	1	0.814698	0.833228
0	0.01853	0.016677	1	0.833228	0.849905
0	0.016677	0.015009	1	0.849905	0.864915
0	0.015009	0.013509	1	0.864915	0.878423
0	0.013509	0.012158	1	0.878423	0.890581

Notice from the table above and from the graphs in **Figure 31** that, due to the larger feedback coefficient, the impulse response decreases more slowly. In addition, notice that the smaller feedforward coefficient means the step response climbs more slowly toward the input value of 1.

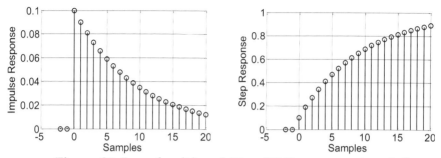

Figure 31 Impulse (L) and Step (R) Responses ($\alpha = 0.1$)

Note that 20 samples is not a magic number for plotting impulse and step responses; it is just the number chosen for these examples. In general, one needs only plot the portion of interest, be it a few or a few thousand samples. Although IIR filters technically have infinite responses, one usually finds the most interesting portion of the response in a fairly small number of samples.

One more item of interest about impulse responses: the impulse response is often denoted $h(n)$, and it turns out to be the inverse z-transform of the frequency response $H(z)$. However, as promised previously, for practical filtering applications, computation of the z-transform is rarely necessary.

1.1.7.2.3 Noise Characteristics

Digital filters are prone to several different types of disturbances arising largely due to the finite precision involved in their implementation. It is probably not fair to call all of these disturbances *noise*, but that might be the most recognizable general name for the effects.

Note that the effects discussed herein are focused on fixed-point filter implementations. Floating point implementations are not completely immune to these issues, since they still have a finite number of bits. But typically these issues will be much less pronounced. Furthermore, even for fixed point implementations with twenty-four or more bits, many of these problems are generally negligible or at least easily manageable.

There are primarily two ways that quantization effects can cause issues for digital filters. One is coefficient quantization, the other is quantization of the actual filter arithmetic. Let us first discuss coefficient quantization.

Typically a filter is designed and analyzed using a high precision tool such as Matlab or some other computer software. However, in implementation, the coefficients are typically quantized to some lower precision. This need not be a catastrophe, but it must be done with care to avoid issues.

Consider a 10 Hz highpass filter with Q = 2 designed for operation at a sample rate of 48 kHz. We will learn to design this filter in the **Handbook** section of this book. In the meantime, the coefficients are:
B = [0.999643429999579 -1.999286859999158 0.999643429999579]
A = [1.000000000000000 -1.999285845392226 0.999287874606090]

The magnitude of each of the two poles is 0.999643873890142, so this filter is stable; the magnitude response is the solid line in **Figure 32**.

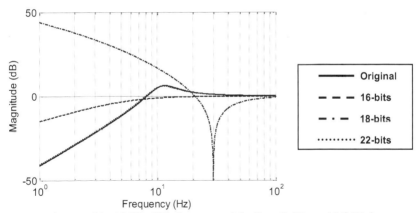

Figure 32 10 Hz Highpass with Q = 2 (Fs = 48 kHz)

Now, let us quantize these coefficients to a few different levels. First, 16-bits:
B = [0.9996337890625 -1.999267578125 0.9996337890625]
A = [1.00000000000000 -1.999267578125 0.999267578125]

Now the magnitudes of the two poles are no longer the same; in fact, one of them is 1, making this filter only marginally stable. The magnitude of the other pole is 0.999267578125.

The 16-bit coefficients give rise to the dashed line in the figure, which only slightly resembles the original magnitude response.

Is 18-bit quantization any better?
B = [0.999649047851563 -1.999282836914063   0.999649047851563]
A = [1.000000000000000 -1.999282836914063   0.999282836914063]

Magnitudes of poles: 1.000000000000000, 0.999282836914063.

This one gives rise to the dash-dotted line in the figure, which is nothing at all like the desired response. However, at 22-bits things straighten out significantly:
B = [0.999643325805664 -1.999286651611328   0.999643325805664]
A = [1.000000000000000 -1.999285697937012   0.999287605285645]

The pole magnitudes are both 0.999643739181937. This gives rise to the dotted line that tracks very closely to the original solid line.

Also note that there are different methods of quantizing, which will create slightly better or worse results for different situations.

The other primary quantization issue is the generation of noise itself. Think of it this way: every time two numbers are multiplied, the precision required to perfectly maintain the result increases. However, since the machine precision does not increase, at some point, the result must be quantized back to machine precision. A long FIR filter may have hundreds or thousands of multiplies. If a quantization step follows each of these, the resulting noise can be noticeable for some applications. Where possible, it is preferable to accumulate the products into a large accumulator and keep the quantization steps to a minimum.

In IIR filters, there are typically far fewer multiplies, but, as seen below, the filter itself can amplify any quantization noise created.

Figure 33 shows a second-order Direct Form I filter, with annotations to illustrate the fixed point problem. Note that the filter has an n-bit datapath and m-bit coefficients. This means that, without quantization, their product will need n+m bits. Of course the system has only n bits, so we must quantize at some point. It is preferable to quantize only once after all the products have been accumulated into an adequately large accumulator at the location of the sum. Many modern processing structures allow this.

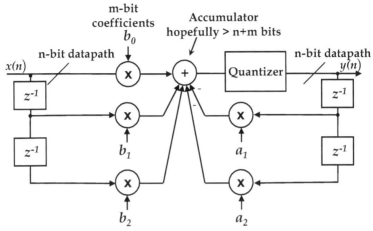

Figure 33 2nd-order Filter with Quantizer Shown

Nevertheless, quantizing even once per cycle still creates quantization noise, though it will be small for systems with adequate precision. And, for IIR filters, the problem can be even worse since the filter itself can amplify the noise. For example, consider **Figure 34**.

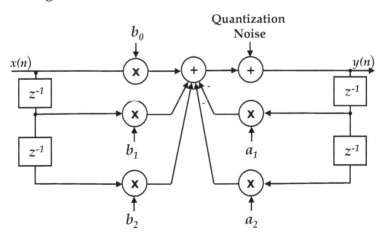

Figure 34 2nd-order Filter with Quantization Noise Model Shown

Here we have replaced the quantizer with a noise source. This is the model used to analyze quantization noise. One such analysis is computation of what is known as the noise transfer function. In this case, we turn off the regular filter input, $x(n)$, and compute a transfer function between the quantization noise input and the regular filter output, $y(n)$. The noise transfer function for this case is shown in **Equation 25**.

$$NTF(z) = \frac{\sigma_Q}{1 + a_1 z^{-1} + a_2 z^{-2}} \qquad \text{Equation 25}$$

Noise Transfer Function for a 2nd-order DFI IIR Filter

Here the σ_Q is the standard deviation of the quantization noise of the filter, which will vary somewhat due to the way the quantization is done. But, for every value of σ_Q, the noise transfer function will be governed by this denominator.

Figure 35 is a plot of the noise transfer function (NTF) of our 10 Hz highpass filter from the example above, with $\sigma_Q = 1$. This is what will act upon whatever level of quantization is present. The 120 dB peak of this plot is about 20-bits of amplification. That is, if the quantization is at 32-bits, this filter effectively moves the quantization noise level to 12-bits down, at least in the active region of the filter. This example was chosen to show a particularly troublesome filter. Let us try a more typical example.

Figure 36 is the noise transfer function of our 2 kHz Butterworth lowpass filter example from above. In this case notice that the effect is really quite benign.

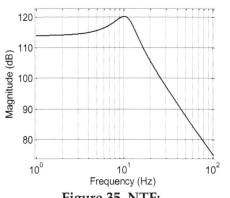
Figure 35 NTF:
10 Hz HP with Q = 2, Fs: 48 kHz

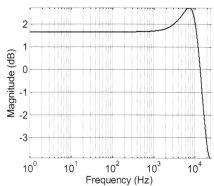
Figure 36 NTF: 2 kHz Butterworth Lowpass (Fs = 10 kHz)

Chapter 1 Introduction & Theory

There is only a slight boost at lower frequencies, and then a noise attenuation at higher frequencies. In other words, one need not be terrified of the noise generated by digital filters; it is usually manageable. But one does need to be aware that this can be a problem, and know where to look should it occur. Note that poles closer to the unit circle, such as in the case of a very low cutoff frequency and a high sampling rate, are potential issues and should be the first place to investigate if a problem arises. For more on noise analysis, please see **Reference 15** pages 181 – 216.

Now for a few words on limit cycles. We discussed how poles on the unit circle create an oscillator. Therefore, poles of magnitude 1 should be avoided unless an oscillator is the desired result. However, there is yet another way that IIR filters can potentially oscillate. In this case, the nonlinear effects of quantization can cause oscillatory patterns in the filter output. Usually these have small magnitude and may not be an issue for certain applications. However, in some applications, such as audio filters, limit cycles may be undesirable since they can create audible tones within the audio band. By way of example of a limit cycle, consider the filter below [**Ref. 6** page 345]:

$$y(n) = x(n) - 0.9y(n-1)$$

This is similar to our alpha filter and has its real pole at -0.9, which means the filter is strictly stable. However, let us now assume that this filter is given an input of 10 for n=0 followed by 0s for all other time steps (this is an impulse scaled by 10). If this filter is implemented with full precision, the energy of the output diminishes with time. However, if instead the output is rounded to an integer value, the filter gets into a limit cycle.

Rather than rounding if we use magnitude truncation, which means that we always truncate in such a way as to make the magnitude smaller, independent of sign (MT(-1.7) = -1, MT(1.7) = 1), the limit cycle can be suppressed. Consider the table below, where y is the full precision output, y_R is the output for the case where rounding is used, and y_{MT} is the output where magnitude truncation is used.

Note that both the rounding and the magnitude truncation cases exhibit quantization effects: their outputs are different than the full precision case. But the magnitude truncation case does not oscillate, and exhibits overall similar behavior to the full precision filter.

n	x(n)	y(n-1)	y(n)	y_R(n-1)	y_R(n)	y_{MT}(n-1)	y_{MT}(n)
0	10	0	10	0	10	0	10
1	0	10	-9	10	-9	10	-9
2	0	-9	8.1	-9	8	-9	8
3	0	8.1	-7.29	8	-7	8	-7
4	0	-7.29	6.561	-7	6	-7	6
5	0	6.561	-5.9049	6	-5	6	-5
6	0	-5.9049	5.31441	-5	5	-5	4
7	0	5.31441	-4.78297	5	-5	4	-3
8	0	-4.78297	4.304672	-5	5	-3	2
9	0	4.304672	-3.8742	5	-5	2	-1
10	0	-3.8742	3.486784	-5	5	-1	0
11	0	3.486784	-3.13811	5	-5	0	0
...							...
49	0	0.063627	-0.05726	5	-5	0	0
50	0	-0.05726	0.051538	-5	5	0	0

While magnitude truncation is not guaranteed to eliminate all limit cycles in all filters, it is generally a powerful tool against them.

1.1.8 Filter Transforms

Often filter transforms are the very heart of filter design methodologies. However, to simplify the design process, the **Handbook** section of this book gives straightforward design equations, eliminating the steps of first creating a prototype design, and then transforming it to achieve the final design. Instead, any necessary transforms are built into the equations in that section. However, being aware of the transforms familiarizes the reader with what is behind the scenes and enables going beyond the formulas presented in this book, should the need arise.

There are primarily three types of transforms of interest: transforms for converting analog lowpass filters to other filter types, transforms for converting analog filters to digital filters, and transforms for converting a digital filter of one frequency to a different frequency. In this section we shall address only the first two types, since we will use them in the **Handbook** section. The third type is addressed in the **Advanced Topics** section of this book.

Before there was digital, analog filters ruled the world. Even as digital systems became commonplace, it was natural to design digital replicas of the myriad known and understood analog filters. Even today, many filters begin life in analog, and are then transformed into digital.

Filters often begin as analog lowpass filters, are transformed into highpass, bandpass or bandstop filters, and are then transformed to digital. Therefore, before discussing transforms for converting such filters from analog to digital, we shall introduce the transforms for converting analog lowpass filters to the other filter types.

LP to LP $\qquad s \to \dfrac{s}{\omega_c}$ $\qquad$ **Equation 26**

LP to HP $\qquad s \to \dfrac{\omega_c}{s}$ $\qquad$ **Equation 27**

LP to BP $\qquad s \to \dfrac{1}{BW}\left(s + \dfrac{\omega_c^2}{s}\right)$ $\qquad$ **Equation 28**

LP to BS $\qquad s \to \dfrac{BWs}{s^2 + \omega_c^2}$ $\qquad$ **Equation 29**

Analog Filter Transforms

The first transform, **Equation 26**, converts a lowpass filter to lowpass at a different cutoff frequency. This is used for a prototype filter which is designed for a cutoff of 1 radian per second and which the designer wishes to move to a different cutoff.

For example, consider a first-order analog Butterworth lowpass filter with cutoff of 1 rad/s:

$$H(s) = \dfrac{1}{s+1} \qquad \text{Equation 30}$$

First-order Analog Butterworth Filter

The frequency response of this filter is shown in **Figure 37**.

Now, if we wish to use **Equation 26** to move the cutoff to 10 radians per second, we substitute $s/10$ for s and normalize:

$$\frac{1}{\frac{s}{10}+1} = \frac{10}{s+10}$$

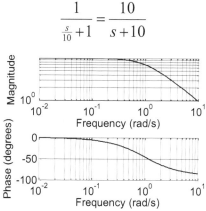

Figure 37 First-order Analog Butterworth LP $\omega_c = 1$ rad/s

The frequency response, plotted in **Figure 38**, demonstrates the success of the transformation (compare the frequency axes).

Now, instead of a lowpass filter with a 10 rad/s cutoff, let us make it a highpass with the same cutoff: $\dfrac{1}{\frac{10}{s}+1} = \dfrac{s}{s+10}$

The new frequency response is plotted in **Figure 39**.

A bandpass filter has two cutoff frequencies f_L and f_H (low and high) and the bandwidth is defined as the frequency range between them $BW = f_H - f_L$. The center frequency is defined as: $f_c = \sqrt{f_L \cdot f_H}$.

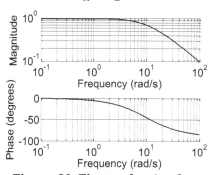
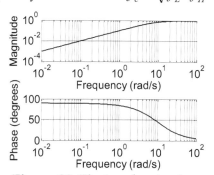

Figure 38 First-order Analog Butterworth LP $\omega_c = 10$ rad/s

Figure 39 First-order Analog Butterworth HP $\omega_c = 10$ rad/s

Now, to create a bandpass with cutoffs at 10 and 40 rad/s, the bandwidth (BW) is 30 rad/s and the center frequency is 20 rad/s.

This transform doubles the filter order so the first-order lowpass becomes a second-order bandpass.

$$\frac{1}{\frac{1}{BW}\left(s+\frac{\omega_c^2}{s}\right)+1} = \frac{BW}{s+\frac{\omega_c^2}{s}+BW} = \frac{BWs}{s^2+BWs+\omega_c^2} \rightarrow \frac{30s}{s^2+30s+400}$$

The frequency response of **Figure 40** clearly shows the desired result.

Finally, let us derive a bandstop filter with the same specifications as our bandpass. This transform also doubles the filter order:

$$\frac{1}{\frac{BWs}{s^2+\omega_c^2}+1} = \frac{s^2+\omega_c^2}{s^2+BWs+\omega_c^2} \rightarrow \frac{s^2+400}{s^2+30s+400}$$

The desired response is shown in **Figure 41**. The discontinuity at 20 rad/s in the phase plot is the oft-seen *phase wrapping*. The right part of the graph could equivalently be shifted down and attached to the left half to show no discontinuity and a transition between 0 and -180 degrees. Some software tools, such as Matlab, have unwrapping commands (UNWRAP) to remove these discontinuities, but it is also common to plot them as is. Note that the previous plots did not wrap because their phase responses were never greater than 90 degrees, where the wrapping occurred.

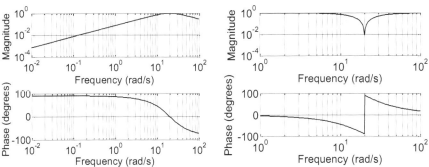

Figure 40 2nd-order Analog Butterworth BP ω_c=20, BW=30 rad/s

Figure 41 2nd-order Analog Butterworth BS ω_c=20, BW= 30 rd/s

Now that we have learned to convert analog lowpass filters into various other filter types, we shall turn our attention to methods of converting analog filters to digital. A very common method is the bilinear transform, which is given by **Equation 31** below.

$$H(z) = H(s)\Big|_{s=\frac{k(z-1)}{z+1}}$$

$$k = 2F_s \quad \text{or} \quad k = \frac{2\pi f_M}{\tan\left(\frac{\pi f_M}{F_s}\right)} \quad \text{to match at } f_M$$

Equation 31

Bilinear Transform

In this equation, k may simply take on the value of $2F_s$, but the alternate equation for k does a frequency warping that will create an exact match between the analog and digital filters at frequency f_M. Using the simple definition of k, let us convert our 10 radian per second lowpass from above to a digital filter that is operational at a 100 Hz sample rate:

$$\frac{10}{s+10} \rightarrow \frac{10}{\frac{2F_s(z-1)}{z+1}+10} = \frac{10z+10}{2F_s(z-1)+10z+10} = \frac{10z+10}{200(z-1)+10z+10}$$

$$= \frac{10z+10}{200(z-1)+10z+10} = \frac{10z+10}{200z-200+10z+10} = \frac{10z+10}{210z-190}$$

$$H(z) = \frac{\frac{10}{210} + \frac{10z^{-1}}{210}}{1 - \frac{190z^{-1}}{210}}$$

The final transfer function is put into implementation form and normalized such that the first denominator coefficient is 1.

In **Figure 42** the resulting filter is plotted using a dotted line against the solid line of the original analog filter. Note that the agreement is very good although there is a slight deviation at the far right edge. This is typical and usually acceptable, though the deviations can sometimes be pronounced near the maximum frequency of $F_s/2$.

Although we used the simple definition for k in this example, the equations in the **Handbook** section actually use the more complicated version. This was done for three reasons: 1) this definition is commonly used, so the resulting filters are more likely to be similar to those designed using filtering tools, 2) as used in the **Handbook**, this forces the digital filter to match its analog prototype at the critical frequency, and 3) the factor π cancels other πs leading to equations that are simpler overall, although they do require computing a tangent.

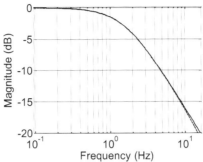

Figure 42 Digital and Analog Butterworth LP

Another popular analog to digital conversion method, the impulse invariant transform, is accomplished as follows:

1. Factor the denominator polynomial and write $H(s)$ as a partial fraction expansion: $H(s) = \sum_{i=1}^{N} \frac{C_i}{s - P_i}$ where P_i are the poles of the analog filter and C_i are the numerator coefficients arising from the partial fraction expansion.

2. The digital approximation is given by this equation:

$$H(z) = \frac{1}{F_s} \sum_{i=1}^{N} \frac{C_i z}{z - e^{P_i/F_s}}$$

Equation 32

Impulse Invariant Transform

Let us try this method for our bandpass filter example from above, this time using a sample rate of 2 kHz. Note that for the equation $\frac{30s}{s^2 + 30s + 400}$ the poles are at s = -14.999999999999998 ±13.228756555322953i. (The stability rules are different for analog filters, and this one is stable.)

Next, we must do a partial fraction expansion:

$$\frac{30s}{s^2 + 30s + 400} = \frac{C_1}{s + 14.999999999999998 - 13.228756555322953i} +$$

$$\frac{C_2}{s+14.999999999999998+13.228756555322953i}$$

After some arithmetic, the values C_i turn out to be:

$$C_1 = 15.0+17.008401285415225i \quad C_2 = 15.0-17.008401285415225i$$

Plugging in to **Equation 32**:

$$H(z) = \frac{1}{F_s}\sum_{i=1}^{N}\frac{C_i z}{z-e^{P_i/F_s}} = \frac{1}{2000}\left(\frac{(15.0+17.008401285415225i)z}{z-e^{(-14.999999999999998+13.228756555322953i)/2000}}\right.$$

$$\left.+\frac{(15.0-17.008401285415225i)z}{z-e^{(-14.999999999999998-13.228756555322953i)/2000}}\right)$$

$$=\frac{1}{2000}\left(\frac{(15.0+17.008401285415225i)z}{z-0.992506343347096+0.006564908136399i}\right.$$

$$\left.+\frac{(15.0-17.008401285415225i)z}{z-0.992506343347096-0.006564908136399i}\right)$$

$$=\frac{1}{2000}\frac{30z^2 - 29.457531728867796z}{z^2-1.985012686694192z+0.985111939603063}$$

$$=\frac{0.015z^2-0.014999253742192z}{z^2-1.985012686694192z+0.985111939603063}$$

The result is plotted in **Figure 43**. Although the digital filter is shown with a dotted line and the analog filter is solid, it is difficult to distinguish the two. Of course, the match will not always be this good, depending upon sample rate, and the filter itself. Note that our 30 rad/s bandwidth is equal to $30/(2\pi) \approx 4.8$ Hz as seen in the plot by taking the difference between the two -3 dB points.

Notice that each of these two transform methods produces filters in analysis form; we multiplied by $\frac{z^{-n}}{z^{-n}}$ to put them in the implementation form.

There exist many ways to transform analog filters to digital, but for most filters, one of these two methods will suffice. For more information on transforms, please see **Reference 5** pages 168 – 190.

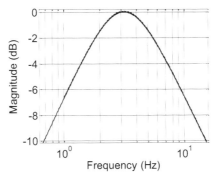

Figure 43 Digital and Analog Butterworth BP

1.1.9 Filter Theory Summary and Quick Reference

Digital filters manipulate numbers to create frequency-based signal effects. Filters are typically characterized by the frequencies they pass: lowpass filters have low frequency passbands and high frequency stopbands; highpass filters have the opposite. Bandpass filters pass a band of frequencies and attenuate frequencies outside that band; bandstop filters attenuate only a certain band. An allpass filter attenuates no frequencies, but typically modifies the phase in some way. A filter with mirror image numerator and denominator coefficients will be allpass.

The region between the stopband and passband is called the transition band. Typically the -3 dB point is called the corner or critical frequency and is denoted f_c. For a bandpass or bandstop filter, f_c denotes the center frequency.

There exist numerous structures for implementing digital filters. We traditionally call the input of a filter $x(n)$ and the output $y(n)$, where n begins at 0, and $x(n - m)$ indicates a sample from m steps ago. The z^{-1} in these filter structures is the unit delay, a storage element. Passing $y(n)$ though one of these unit delays gives us $y(n-1)$. If the filter has feedback from the output, it is IIR, otherwise it is FIR. Be careful, however, since the feedback can be difficult to spot in some filter structures.

A simple average is a type of finite impulse response (FIR) filter. Another type of averaging filter is a first-order infinite impulse response or IIR filter, although it does not create an average in the usual sense.

Filters are defined using difference equations with coefficients applied to present and past values of the input signal, and past values of the output signal.

The impulse response of a filter is the output of the filter when an impulse, 1 followed by 0s, is input, assuming all filter nodes were initially zero. The impulse response of an FIR filter is just the filter coefficients. The step response is the response to a step function: 1 starting at the first sample and 1s forevermore.

Difference equations are converted to z-transform notation using the following rule: $g(n-\Delta) \rightarrow G(z)z^{-\Delta}$. This allows us to find the filter transfer function: $H(z) = \dfrac{Y(z)}{X(z)}$. In implementation form, this will be a function of z^{-1}. To plot this function we substitute $z = e^{i\omega}$ (or $z^{-1} = e^{-i\omega}$), $\omega = \dfrac{2\pi f}{F_s}$ and use software that can manipulate complex numbers to plot the magnitude.

Phase: $P(\omega) = \arctan 2(\operatorname{Im}[H(\omega)], \operatorname{Re}[H(\omega)])$

Group delay: $D_G(\omega) = \dfrac{-dP(\omega)}{d\omega}$

The order of a filter is the value of the largest delay in the difference equation, or the largest power of z in the transfer function.

To check stability, we convert to the analysis form by multiplying both numerator and denominator by z^n, where n is the order. Since we multiply by the same thing both top and bottom, we are multiplying by one and not disturbing the equality. In analysis form we can find the zeros, which are the roots of our numerator equation, and the poles, which are the roots of the denominator equation. For stability the magnitudes of the poles must be strictly less than 1.

The angles of poles and zeros are related to the active frequencies of the filter. The filter's frequency band, between 0 Hz and $F_s / 2$, maps to the region between 0 and π on the unit circle.

Filters with real coefficients (all filters in this book) have poles and zeros that are either real or appear in complex conjugate pairs. The complex conjugate has the same real and imaginary values except that the sign of the imaginary part is reversed. In the complex plane, conjugate pairs are mirror images about the real axis.

Coefficient quantization can cause the transfer function to change unacceptably, and even cause a stable filter to go unstable. But it is easy to check for these issues and, therefore, to avoid them.

Quantization of the filter arithmetic can cause limit cycles and also generates noise. The noise transfer function relates the quantization noise created inside the filter to the noise appearing on the output. For a Direct Form I second order IIR filter it has this form:

$$NTF(z) = \frac{\sigma_Q}{1 + a_1 z^{-1} + a_2 z^{-2}}$$

Finally, there are many ways to change an analog filter to digital. One of the most popular is the bilinear transform:

$$H(z) = H(s)\Big|_{s=\frac{k(z-1)}{z+1}} \text{ where } k = 2F_s \text{ or } k = \frac{2\pi f_M}{\tan\left(\frac{\pi f_M}{F_s}\right)}$$

Chapter 2: Digital Filter Applications

In a previous section we discussed various filter types. A figure from that section has been repeated as **Figure 44** below. We will discuss applications for each of these filter types.

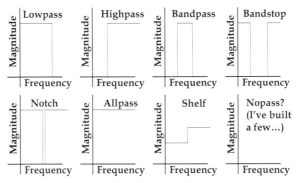

Figure 44 Types of Filters

Noise removal is a typical use for filters. When it is known that the information of interest is within a certain frequency range, a filter can maintain needed frequencies while attenuating others.

The **Handbook** section of this book gives design equations for numerous varieties of each of these different filter types. For variety, and to familiarize the reader with the various types of filters, a number of different filters are used in these examples. In practice, the user must trade off the characteristics of the different filter types for the application at hand. If a fast cutoff is needed, higher order filters will help, but for any given order, the Chebychev filters generally cut off faster than the others. On the other hand, more aggressive filters tend to have longer time responses. This is one of the tradeoffs to be made.

Generally speaking, Butterworth filters are a good first choice. They are a good balance between being well behaved and having nice roll off characteristics. From that point, the user can move on to different filters as needed: Chebychev for faster roll offs; Bessel for nicer phase characteristics; Linkwitz-Reilly for audio crossovers, and so forth. You will definitely not always make the best choice the first time, but you can always test your choices and make new ones if the performance is not as desired. With practice you'll get better at making appropriate initial selections.

Let us now define a test signal to use in some of our examples below. Consider a 937.5 Hz sinusoid sampled at 48 kHz. A few cycles of such a signal, along with the associated spectrum, computed over a longer sample, are plotted in **Figure 45**. You might have wondered why this particular frequency was chosen. Part of the answer is that a fast Fourier transform, or FFT, was used to compute the spectrum. Typically FFTs are performed on data of length 2^n. While there are other options, the frequency chosen allows an integer number of sine cycles within such sample sizes. Sometimes called a synchronous signal, this prevents artifacts caused by violating the FFT's implicit assumption that the signal within the FFT will continue to repeat indefinitely.

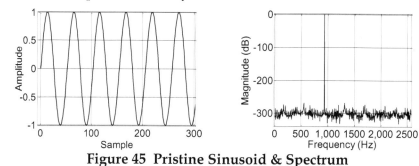

Figure 45 Pristine Sinusoid & Spectrum

This sinusoid is pristine. The noise floor seen at -300 dB is just from numerical precision. When you consider that -300 dB is a factor of $10^{(-300/20)} = 10^{-15}$, which is fourteen zeroes to the right of the decimal followed by a one, it's clear that this is not much noise at all!

Now, since we want to show the noise mitigation capabilities of filters, we'll add some noise to this signal, and then pass it through some filters to clean it up. **Figure 46** shows our sinusoid from above but now with random noise added to it. It's difficult to be sure the sine is still in there when looking at the sample domain plot at left. The FFT is still able to find the sine, but notice that the noise floor is now only 30 dB down, as opposed to almost 300 dB down for the pristine sine. Clearly there is significant degradation in the signal.

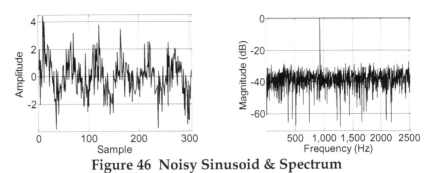

Figure 46 Noisy Sinusoid & Spectrum

Various filters could be used to mitigate the noise on our sinusoidal signal. Let's begin by applying a fourth-order type I Chebychev bandpass filter centered at 937.5 Hz, with a bandwidth of 5 Hz, a passband ripple of 0.5 dB and, of course, computed for the 48000 Hz sample rate used for our signal. The **Handbook** section gives the equations for computing the coefficients for this filter, which turn out to be as shown below:

$b_0 = 0.00000016148462074$ $a_0 = 1$
$b_1 = 0$ $a_1 = -3.968994197798568$
$b_2 = -0.00000032296924148$ $a_2 = 5.937298216671415$
$b_3 = 0$ $a_3 = -3.967147598767473$
$b_4 = 0.00000016148462074$ $a_4 = 0.999069704227281$

The frequency response of this filter is shown on the left side of **Figure 47** while the filter output is shown at right. Note that in practice, we might not always know the exact frequency of the signal we seek and therefore could not use such a narrow filter to find it. But when we do know the exact frequency, a narrow filter gives us an opportunity to mitigate more noise, as we shall see.

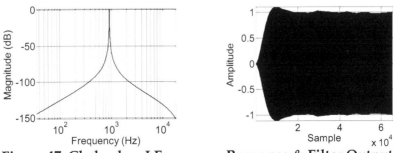

Figure 47 Chebychev I Frequency Response & Filter Output

The filter output is sinusoidal, as we shall see below, but since we have plotted tens of thousands of samples the individual sinusoidal excursions are not distinguishable. That's OK because we are initially interested in the envelope of this signal. Notice how it climbs from zero and overshoots somewhat then never really reaches a steady state. The reasons for this have to do with the time domain characteristics of the filter. The impulse response is plotted in **Figure 48** below. Technically the impulse response is infinite in length, but it gets much less interesting after the first twenty thousand samples which are plotted at left. As with the filter output itself, there are so many samples plotted that we can only see the envelope, so we have also plotted the first forty samples at right. The response builds in this sine-like fashion creating the envelope at left.

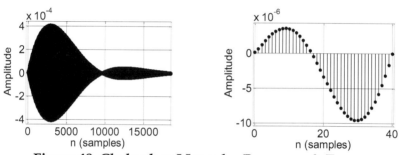

Figure 48 Chebychev I Impulse Response & Zoom

The step response of the filter is shown in **Figure 49**. Although we are seldom filtering an impulse or a step function, the impulse and step responses give us an idea what will happen when we put a signal through the filter. We see that it takes the filter a while to build up to the level of the input signal, as we saw in the filter output as well. This is sometimes referred to colloquially as the *charge time* of the filter. This means that we should not rely upon the output of the filter until it reaches something close to a steady state response. Of course we noticed that the envelope of our filter output is wavy even after tens of thousands of samples. This is the result of the large noise spikes in the input signal. If the output were far less noisy, such waves may not be detectable. Of course they would likely be even larger for heavier noise.

Chapter 2 Digital Filter Applications

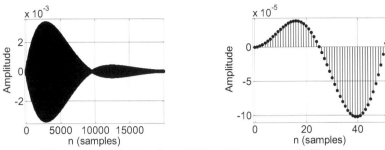

Figure 49 Chebychev I Step Response & Zoom

One of the reasons for considering the time domain responses of filters is so that we recognize the tradeoffs we are making when we apply certain filters. We have applied a very aggressive filter here and it certainly mitigates the noise nicely. But in practice we might choose a filter with a shorter time response even if it doesn't quite have the noise mitigation capability. We discuss other options below.

A few cycles of the filter output and the associated spectrum are shown in **Figure 50**. We took these output samples in a stable portion of the overall response, and we did not use the beginning of the response, where the output is erratic, in the spectrum calculation.

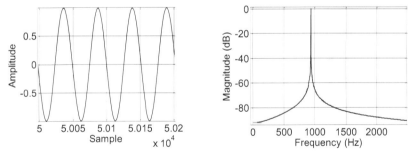

Figure 50 Noise Mitigated Sinusoid & Spectrum

Notice in **Figure 50** that the resulting spectrum is the original sine plus the noise all convolved with the filter; one can see the filter spectrum widening the base of the sine spike, as would be expected.

In many instances we might not know the frequency of interest exactly, but might rather know that it will occur within a particular region. In the case of our example, we might know that the signal of interest is in the region 500 Hz to 1500 Hz. In this case, we need a wider bandpass. Just for variety, let's now apply a fourth order Bessel bandpass centered at 1000 Hz and with a 1000 Hz bandwidth. This filter has the coefficients shown below:

$b_0 = 0.010572698119032$ $a_0 = 1$
$b_1 = 0$ $a_1 = -3.602619995730391$
$b_2 = -0.021145396238065$ $a_2 = 4.887023654151077$
$b_3 = 0$ $a_3 = -2.960159073968458$
$b_4 = 0.010572698119032$ $a_4 = 0.675997654134948$

The frequency response of this filter is shown in **Figure 51** below next to the output signal. Not only does this filter not mitigate the noise at all in the passband, it doesn't drop as far as quickly as did the Chebychev filter above. Not surprisingly, the output signal is much noisier than was the one above. However, notice also that it doesn't appear to take as long for the output amplitude to rise as it did in the Chebychev.

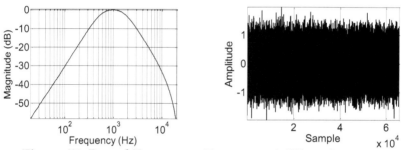

Figure 51 Bessel Frequency Response & Filter Output

Considering the impulse and step responses, in **Figure 52** we see that the active regions are not nearly as long as those for the Chebychev. This is one of the tradeoffs between the two filters.

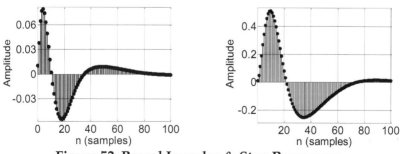

Figure 52 Bessel Impulse & Step Responses

Figure 53 shows a few cycles of the noise mitigated sine wave and the resulting spectrum.

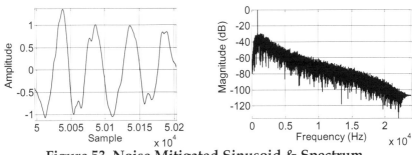

Figure 53 Noise Mitigated Sinusoid & Spectrum

As expected, these exhibit more noise than those for the Chebychev filter above, but also still exhibit significant improvement over the original noisy signal.

In some applications we might not have the luxury of needing only a small passband, but we might still know we need no information above a certain frequency. In such a case, we could consider applying a lowpass filter that cuts off beyond the upper edge of the care-about region. For our example case, let us try a second-order Butterworth lowpass with a cutoff of 5500 Hz, the coefficients for which are shown below:

$b_0 = 0.084625382257444$ $a_0 = 1$
$b_1 = 0.169250764514888$ $a_1 = -1.025542903685195$
$b_2 = 0.084625382257444$ $a_2 = 0.364044432714971$

The frequency response and filter output are shown in **Figure 54** followed by the impulse and step responses in **Figure 55**. Notice how the (active regions of the) impulse and step responses are the shortest we've seen. It is to be expected that they will be less active for lower order and less aggressive filters.

Figure 56 shows a few cycles of the noise mitigated signal and the resulting spectrum. Clearly more noise remains than in either of the previous examples, but there is also still a clear improvement over the original noisy signal.

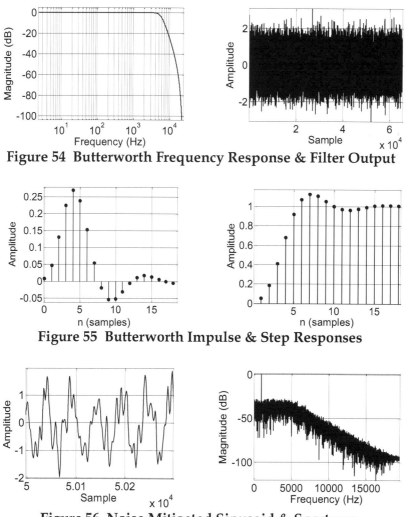

Figure 54 Butterworth Frequency Response & Filter Output

Figure 55 Butterworth Impulse & Step Responses

Figure 56 Noise Mitigated Sinusoid & Spectrum

Of course we have not exhausted the subject of noise mitigation, but you get the point that there are numerous ways to attack, depending upon the exact circumstances.

Let us now move on to a different problem: offset removal. For this problem, let us take our sinusoid from above, and add an offset to it.

The result is plotted in **Figure 57**. Notice the energy near 1000 Hz, our original sine, and the energy at 0 Hz, or DC, which is the offset.

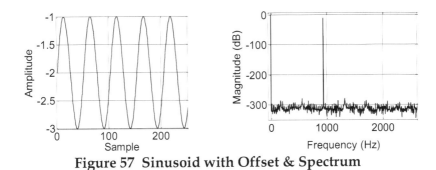

Figure 57 Sinusoid with Offset & Spectrum

An excellent tool for offset removal is the highpass filter. Here we primarily worry about how much of the lowest frequency information we can do without, and make the corner below that point. For this example, let's use a third-order Chebychev II highpass, with the coefficients shown below:

$b_0 = 0.992376404640583$ $\qquad a_0 = 1$
$b_1 = -2.977129162909445$ $\qquad a_1 = -2.984694416479490$
$b_2 = 2.977129162909445$ $\qquad a_2 = 2.969505790133198$
$b_3 = -0.992376404640583$ $\qquad a_3 = -0.984810928487367$

For this filter, the frequency response and output of which are shown in **Figure 58** below, one specifies the stopband ripple and the corner frequency. The actual 3 dB point may vary due to the ripple setting. For this filter we chose 100 dB of ripple, which is achieved exactly, as can be seen (the "bounce" in the stopband). Though we used 2 Hz for the corner in the design equation, the 3 dB point of this filter ends up being around 60 Hz. If such a high cutoff cannot be tolerated, one can use a lower stopband ripple, or even choose a different filter type that allows setting this independently.

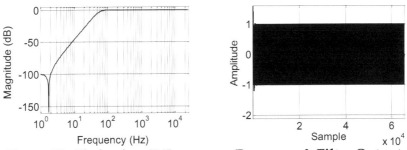

Figure 58 Chebychev II Frequency Response & Filter Output

Although there is some overshooting at the beginning of the output signal, this filter reaches a nice steady state fairly quickly. A few cycles of the offset removed sine and the corresponding spectrum are shown in **Figure 59**. The offset is no longer found in the sample domain, and, correspondingly, the DC energy is greatly attenuated.

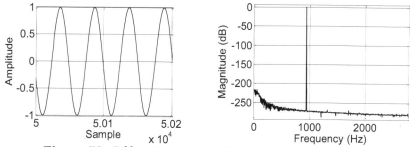

Figure 59 Offset Removed Sinusoid & Spectrum

For demonstrating some other filter types we shall now take our original noise-free sine and add another sine to it. The second sine has the same amplitude, 1, and a frequency of five times the original, or 4687.5 Hz. For fun we add a phase offset to the new sine, so the two do not both go through zero at the same time (on those occasions when the frequencies would otherwise allow). Some samples of the combined signal, along with the spectrum, are shown in **Figure 60**.

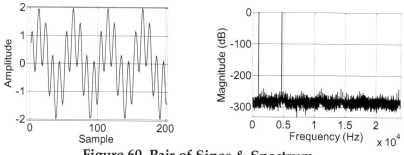

Figure 60 Pair of Sines & Spectrum

Now, if the task is to remove one of the sines, let's say the lower frequency one, we can use a notch filter with center frequency of exactly the sine frequency and a bandwidth of 10 Hz. Notch filters are by definition narrow but the design equations allow control over the bandwidth. The coefficients for this filter are shown below followed by the frequency response and filter output in **Figure 61**.

$b_0 = 0.999345929525232$ $a_0 = 1$
$b_1 = -1.983660766076636$ $a_1 = -1.983660766076636$
$b_2 = 0.999345929525232$ $a_2 = 0.998691859050465$

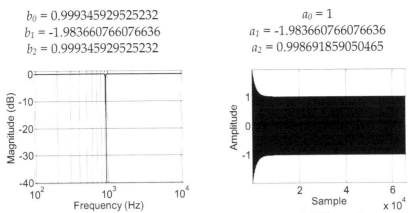

Figure 61 Notch Filter Frequency Response & Filter Output

A few cycles of the output are shown in **Figure 62** along with the resulting spectrum.

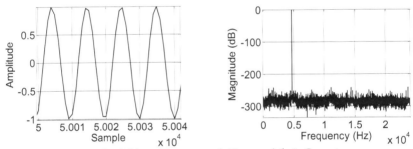

Figure 62 Offset Removed Sinusoid & Spectrum

Now, in circumstances where we do not know the exact frequency of the offending sine, or where there could be a range of interfering frequencies, rather than using the surgical precision of the notch, we might instead choose to use a bandstop filter. For this example, let's add our noise back in, as seen in **Figure 63** below:

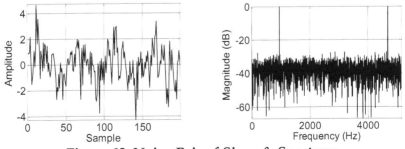

Figure 63 Noisy Pair of Sines & Spectrum

The coefficients for a fourth-order Linkwitz-Reilly bandstop filter centered at 937.5 Hz and with a 500 Hz bandwidth are shown:

$b_0 = 0.937777304662843$ $a_0 = 1$
$b_1 = -3.722899131556044$ $a_1 = -3.844425157260671$
$b_2 = 5.570455806770371$ $a_2 = 5.568457299349689$
$b_3 = -3.722899131556044$ $a_3 = -3.601373105851418$
$b_4 = 0.937777304662843$ $a_4 = 0.877553116746368$

The frequency response of this filter and the filter output are shown in **Figure 64**.

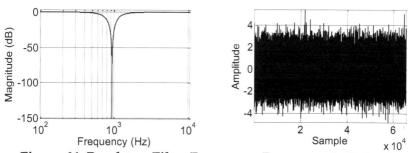

Figure 64 Bandstop Filter Frequency Response & Output

Finally, a small segment of the output is shown alongside the resulting spectrum in **Figure 65**. Of course, the only noise mitigation was in the narrow region where the bandstop filter was applied, so the result is still very noisy, though from the spectrum it is clear that one of the sines has indeed been removed.

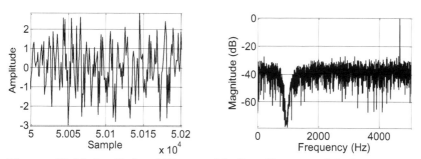

Figure 65 Noisy Pair of Sines with One Removed & Spectrum

Next, to show the functionality of shelf filters, let us consider again the noisy pair of sines of **Figure 63**. To this input signal, let us apply a second order shelf with a gain of 12 dB at high frequencies and a transition at 3 kHz. The coefficients for this filter are shown in the table below while the frequency response and the filter output signal follow in **Figure 66**.

$b_0 = 3.468476463721798$ $a_0 = 1$
$b_1 = -5.937361452868669$ $a_1 = -1.436748664731060$
$b_2 = 2.595796185109824$ $a_2 = 0.563659860694013$

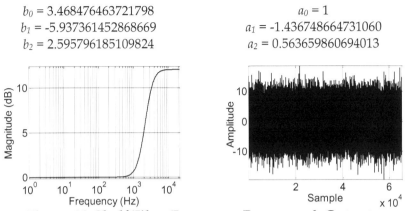

Figure 66 Shelf Filter Frequency Response & Output

Figure 67 shows a number of samples from the filtered signal along with the spectrum. Note that as expected the sines now have a 12 dB magnitude disparity due to the filter. However, something we always must consider when applying a filter is the effect on noise: the high frequency noise floor is also elevated by the application of this filter. (Note that we have renormalized all spectra to a maximum of 0 dB. With respect to the previous plot, this could have been shown with a 0 dB lower frequency sine and a higher frequency sine with a magnitude of +12 dB.)

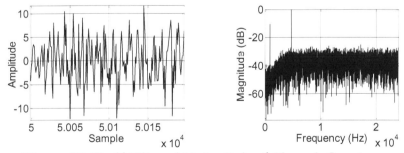

Figure 67 Shelf Filtered Noisy Pair of Sines & Spectrum

A filter type that does not show up in **Figure 44** but which is still important is the bell-shaped equalization or EQ filter. This filter goes to 0 dB at either edge, and then either amplifies or attenuates the frequencies between. Let's consider an application for such a filter. Imagine that you have a sensor with the response shown by the solid line in the right frame of **Figure 68**, and that you would like to use it for a data acquisition that requires 0.5 dB accuracy between 160 Hz and 4 kHz. One way to solve such a problem is to use filters to equalize (sometimes called compensate) the sensor response. Typically this requires filters that can apply some gain or attenuation in a region, such as the EQ filters and shelf filters we have used in other applications.

At left in the figure we show the magnitude responses of three equalization filters with the following parameters (gain, center frequency, and bandwidth): (7 dB, 159 Hz, 34 Hz), (-10.5 dB, 260 Hz, 85 Hz), and (5.1 dB, 4 kHz, 3.5 kHz). At right we show, as a dotted line, the equalized sensor response, achieved by applying the three filters at left, along with the original sensor response (solid).

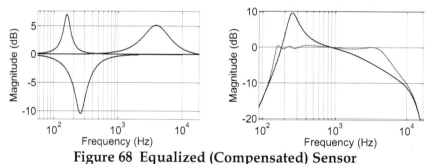

Figure 68 Equalized (Compensated) Sensor

Note that it is technically possible to create the responses of **Figure 68** either by cascading or combining filters in parallel, though the filters in this example are designed to be cascaded. **Section 2.6.5** discusses this topic in more detail.

As a final application example, let us consider an allpass filter. From the standpoint of the way we often think of filters, the very name sounds ridiculous; but allpass filters actually have very useful functions. For one thing, they are sometimes used as building blocks for other types of filters, such as the EQ filters above; but even on their own they are useful, though we must think of them in a little different way than we do other filters.

Consider the frequency response of a fourth-order Butterworth lowpass filter with cutoff at 1.68 kHz, as shown in **Figure 69**.

Although we often plot the magnitude response on a semi-log axis, we have shown it on a linear axis here for two reasons: the roll-off is so steep that it's hard to see anything interesting on a log scale, and the linear axis helps us relate it to the phase response, which is best plotted on a linear axis because we are particularly concerned with phase linearity. And, of course we note here that the phase is not particularly linear.

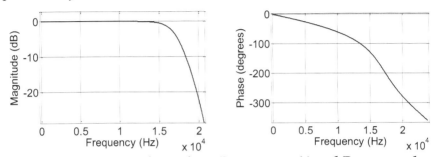

Figure 69 Magnitude & Phase Responses: 4th-ord Butterworth

The best way to understand the effect of phase non-linearity is to consider the group delay as shown in **Figure 70**. Here we see that the amount of delay varies somewhat as a function of frequency.

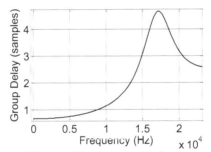

**Figure 70 Group Delay:
4th-order Butterworth**

Now, as group delays and phase non-linearities go, this filter is rather mild. For many, perhaps most applications, no correction is required. In fact, a good rule of thumb in filtering, and signal processing in general, is to do no more than necessary. Not only does it consume processing resources, but the more we handle the data, the more we can damage it through quantization effects, added noise, and so forth. But let us proceed with this example to demonstrate the principle.

The phase responses of three second-order allpass filters are shown at left in **Figure 71**. These filters, the design equations for which are found in the **Handbook** section, have the following critical frequencies and bandwidths: (4600 Hz, 6800 Hz), (8500 Hz, 6800 Hz), (13100 Hz, 6600 Hz). The resulting group delays are shown at right in the figure. Just as we did with magnitude equalization above, these bell-shaped group delays can be used to flatten the overall group delay of a system.

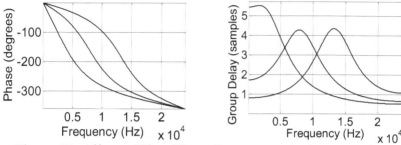

Figure 71 Allpass Filter Phase Responses & Group Delays

Figure 72 shows at left the original phase response of **Figure 69** plotted against a phase response corrected by application of the three allpass filters of **Figure 71**. Note that while the corrected response appears to be more linear, it is also steeper – that's the only direction we can take it. Note also that it does have a bend at high frequencies. This is because there is usually no point in correcting the stopband region of the filter where the signals are highly attenuated anyway. At right the group delays of **Figure 71** are re-plotted against the original group delay of **Figure 70** (shown as a dashed line) and the corrected group delay shown as a dotted line at the top of the plot. We have not corrected the stopband region, and even the passband is not completely flat, though it now varies less.

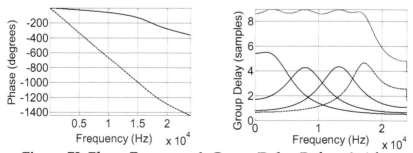

Figure 72 Phase Response & Group Delay Before & After

Part 2: The Handbook

This portion of the book is written to be useful on a standalone basis to facilitate the design of digital filters. The equations herein give straightforward methods to design filters, generally in a single step, without first designing prototypes, and then transforming them as is described in the **Theory** section and often seen elsewhere.

Chapter 3: Infinite Impulse Response (IIR) Filters

As discussed in the **Theory** section, IIR filters have feedback from their outputs which keep their outputs alive even after their inputs have gone away. While FIR filters, discussed in the next chapter, enjoy a great deal of popularity, IIR filters are also extremely useful and are often very efficient to implement. This chapter details the design of a number of IIR filter types. The equations presented in this section directly return filter coefficients from input parameters such as corner frequency, f_c, and sample rate, F_s.

The IIR filters we present have the following transfer function:

$$H(z) = \frac{b_0 + b_1 z^{-1} + b_2 z^{-2} + b_3 z^{-3} + b_4 z^{-4}}{1 + a_1 z^{-1} + a_2 z^{-2} + a_3 z^{-3} + a_4 z^{-4}} \qquad \text{Equation 33}$$

General Equation for IIR Filters up to Fourth Order

Since many of the filters will be of lower than fourth order, some of the coefficients will be 0 (not needed) for those filters.

The equations tend to get more cumbersome as the order increases, so we have chosen to focus only on IIR filters up to fourth-order. Higher order-filters will very often not be required, but, if needed, can be constructed by combining multiple lower-order filters.

2.3.1 Butterworth Filters

You might have spent all afternoon wondering what in the world you'd use a Butterworth filter for. And the answer, in my view, is relatively straightforward: unless I have a specific reason to do otherwise, I start with a Butterworth filter. At times I will discover the reason to do otherwise after first trying the Butterworth. But nine times out of ten the Butterworth is the solution I go with. So, when in doubt, use the Butterworth. Each section below, however, will give you reasons to consider using some of the other options.

This section provides straightforward design equations for a number of different Butterworth digital filters. However, so the process behind these equations will be well understood, we also share pertinent background.

At the heart of Butterworth filters are what are called the *Butterworth Poles*. These deliver a maximally flat characteristic that is widely used in applications. **Equation 34** and **Equation 35** describe these poles in terms of first and second order polynomials. An analog Butterworth lowpass filter with cutoff frequency of 1 rad/s is just $1/B_n(s)$ for whatever order of filter is desired.

For n even:
$$B_n(s) = \prod_{k=1}^{\frac{n}{2}} \left[s^2 - 2s \cos\left(\frac{2k+n-1}{2n}\pi\right) + 1 \right]$$ **Equation 34**

For n odd:
$$B_n(s) = (s+1)\prod_{k=1}^{\frac{n-1}{2}} \left[s^2 - 2s \cos\left(\frac{2k+n-1}{2n}\pi\right) + 1 \right]$$ **Equation 35**

Butterworth Poles

From this point, the desired digital filters are derived through a two-step process: first, the filters are transformed to the analog filter of the desired type – lowpass, highpass, bandpass or bandstop – with the desired cutoff frequency(ies). Second, the resulting analog filters are converted to digital. This is all accomplished using transforms that are detailed in the **Theory** section of this book. In this section we show only the results.

2.3.1.1 Butterworth Lowpass Filters

The following four tables give equations for the coefficients for first- through fourth-order Butterworth lowpass filters. Input the corner frequency and sample rate and compute intermediate variable γ; the coefficients are straightforward from there.

In the first edition of this book I generally showed the denominator polynomial for every coefficient, and also repeated the equations for coefficients with the same values. I did this so that one could refer to the table and come away with the complete equation at a glance. However, it also makes the equations seem more complex than they really are. Generally speaking, one will compute the denominator once, and then divide through by that value. Likewise with the repeated coefficient values. If $b_1 = b_0$, usually we'll just compute b_0 and then set b_1 equal to that value. Therefore, beginning with the second edition I have showed the equations in this simplified form.

First Order Butterworth Lowpass	
Definitions: $\gamma = \tan\left(\pi f_c / F_s\right)$	$D = \gamma + 1 \quad (c_n = c'_n / D)$
Numerator Coefficients	Denominator Coefficient
$b'_0 = \gamma \quad\quad b'_1 = b'_0$	$a'_1 = \gamma - 1$

Note that the b'_0 is only the numerator of the coefficient b_0 and that the relationship is as follows: $b_0 = \dfrac{b'_0}{D}$. In other words, $b_0 = \dfrac{\gamma}{\gamma+1}$, $b_1 = \dfrac{\gamma}{\gamma+1}$, and $a_1 = \dfrac{\gamma-1}{\gamma+1}$. From here on out we'll use the shorthand forms show in the table above.

Second Order Butterworth Lowpass			
Definition: $\gamma = \tan\left(\pi f_c / F_s\right)$		$D = \gamma^2 + \sqrt{2}\gamma + 1 \quad (c_n = c'_n / D)$	
Numerator Coefficients		Denominator Coefficients	
$b'_0 = \gamma^2$	$b'_1 = 2b'_0 \quad b'_2 = b'_0$	$a'_1 = 2(\gamma^2 - 1)$	$a'_2 = \gamma^2 - \sqrt{2}\gamma + 1$

Third Order Butterworth Lowpass	
Definitions: $\gamma = \tan\left(\pi f_c / F_s\right)$	$D = \gamma^3 + 2\gamma^2 + 2\gamma + 1$
Numerator Coefficients	Denominator Coefficients
$b'_0 = \gamma^3 \quad\quad b'_2 = 3b'_0$	$a'_1 = 3\gamma^3 + 2\gamma^2 - 2\gamma - 3$
$b'_1 = 3b'_0 \quad\quad b'_3 = b'_0$	$a'_2 = 3\gamma^3 - 2\gamma^2 - 2\gamma + 3$
$(c_n = c'_n / D)$	$a'_3 = \gamma^3 - 2\gamma^2 + 2\gamma - 1$

Note that the fourth-order case does involve two additional intermediate variables, α and β, but that these are just constants that can be computed exactly using the equation, though a value with adequate precision is also tabled.

Fourth Order Butterworth Lowpass	
Definitions:	
$\alpha = -2\left[\cos\left(5\pi/8\right) + \cos\left(7\pi/8\right)\right] = 2.613125929752753$	
$\beta = 2\left[1 + 2\cos\left(5\pi/8\right)\cdot\cos\left(7\pi/8\right)\right] = 3.414213562373095$	
$\gamma = \tan\left(\pi f_c / F_s\right)$ $\qquad D = \gamma^4 + \alpha\gamma^3 + \beta\gamma^2 + \alpha\gamma + 1$	
Numerator Coefficients	**Denominator Coefficients**
$b'_0 = \gamma^4 \qquad\qquad b'_3 = 4b'_0$	$a'_1 = 2(2\gamma^4 + \alpha\gamma^3 - \alpha\gamma - 2)$
$b'_1 = 4b'_0 \qquad\qquad b_4 = b'_0$	$a'_2 = 2(3\gamma^4 - \beta\gamma^2 + 3)$
$b'_2 = 6b'_0$	$a'_3 = 2(2\gamma^4 - \alpha\gamma^3 + \alpha\gamma - 2)$
$(c_n = c'_n / D)$	$a'_4 = \gamma^4 - \alpha\gamma^3 + \beta\gamma^2 - \alpha\gamma + 1$

Example: f_c = 56 Hz, F_s = 50 kHz. All cases require γ, so we'll compute that first: $\gamma = \tan\left(\pi f_c / F_s\right) = \tan\left(\pi 56 / 50000\right)$, therefore $\gamma = 0.003518598292621$.

Now, for the first order case, $D = \gamma + 1$ = 1.003518598292621, $b'_0 = b'_1 = \gamma$ = 0.003518598292621 and $a'_1 = \gamma - 1$ = -0.996481401707379. The final step is to divide the b'_0, b'_1 & a'_1 each by D to get b_0, b_1 & a_1: $b_0 = b'_0 / D = b'_0 / D = \gamma / D = 0.003518598292621 / 1.003518598292621 =$ 0.003506261168062. b_1 = b_0 and $a_1 = a'_1 / D$ = -0.996481401707379 / 1.003518598292621 = -0.992987477663877.

The other cases are worked out similarly, using the value of γ we computed above. Of course the values of D are different for each case, so those must be computed from the formulas for each case.

Chapter 3 IIR Filters

The fourth order case also includes constants α and β which can be computed from the equations given or the tabled values can be used.

The coefficients for this case are shown in **Table 2**; frequency response plots follow in **Figure 73**.

Notice in the figure that filter magnitude and phase plots are steeper for higher filter orders. This is typical behavior.

1st-Ord	$b_0 = b_1 = 0.003506261168062$ $a_1 = -0.992987477663877$
2nd-Ord	$b_0 = b_2 = 0.123190810717402 \times 10^{-4}$ $b_1 = 0.246381621434805 \times 10^{-4}$ $a_1 = -1.990048023748777$ $a_2 = 0.990097300073064$
3rd-Ord	$b_0 = b_3 = 0.043256647105615 \times 10^{-6}$ $b_1 = b_2 = 0.129769941316846 \times 10^{-6}$ $a_1 = -2.985925693951620$ $a_2 = 2.971950256708682$ $a_3 = -0.986024216703885$
4th-Ord	$b_0 = b_4 = 0.151874765065198 \times 10^{-9}$ $b_2 = 0.911248590391186 \times 10^{-9}$ $b_1 = b_3 = 0.607499060260791 \times 10^{-9}$ $a_1 = -3.981611013431778$ $a_2 = 5.945001929108286$ $a_3 = -3.945169910729457$ $a_4 = 0.981778997482945$

Table 2 Coefficients: Butterworth Lowpass Filters

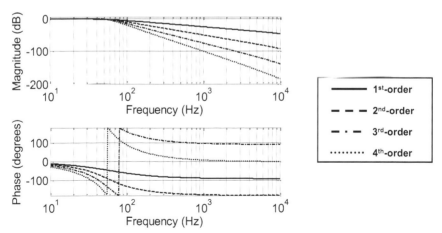

Figure 73 Frequency Responses of Butterworth Lowpass Filters

2.3.1.2 Butterworth Highpass Filters

Similarly, equations for Butterworth highpass filters follow:

First Order Butterworth Highpass	
Definitions: $\gamma = \tan\left(\pi f_c / F_s\right)$	$D = \gamma + 1$ $(c_n = c'_n / D)$
Numerator Coefficients	Denominator Coefficient
$b'_0 = 1$ $b'_1 = -1$	$a'_1 = \gamma - 1$

Second Order Butterworth Highpass				
Definitions: $\gamma = \tan\left(\pi f_c / F_s\right)$			$D = \gamma^2 + \sqrt{2}\gamma + 1$ $(c_n = c'_n / D)$	
Numerator Coefficients			Denominator Coefficients	
$b'_0 = 1$	$b'_1 = -2$	$b'_2 = 1$	$a'_1 = 2(\gamma^2 - 1)$	$a'_2 = \gamma^2 - \sqrt{2}\gamma + 1$

Third Order Butterworth Highpass	
Definitions: $\gamma = \tan\left(\pi f_c / F_s\right)$	$D = \gamma^3 + 2\gamma^2 + 2\gamma + 1$
Numerator Coefficients	Denominator Coefficients
$b'_0 = 1$ $b'_2 = 3$	$a'_1 = 3\gamma^3 + 2\gamma^2 - 2\gamma - 3$
$b'_1 = -3$ $b'_3 = -1$	$a'_2 = 3\gamma^3 - 2\gamma^2 - 2\gamma + 3$
$(c_n = c'_n / D)$	$a'_3 = \gamma^3 - 2\gamma^2 + 2\gamma - 1$

Fourth Order Butterworth Highpass
Definitions:

$$\alpha = -2\left[\cos\left(5\pi/8\right) + \cos\left(7\pi/8\right)\right] = 2.613125929752753$$

$$\beta = 2\left[1 + 2\cos\left(5\pi/8\right) \cdot \cos\left(7\pi/8\right)\right] = 3.414213562373095$$

$$\gamma = \tan\left(\pi f_c / F_s\right) \qquad D = \gamma^4 + \alpha\gamma^3 + \beta\gamma^2 + \alpha\gamma + 1$$

Numerator Coefficients		Denominator Coefficients
$b'_0 = 1$	$b'_3 = -4$	$a'_1 = 2(2\gamma^4 + \alpha\gamma^3 - \alpha\gamma - 2)$
$b'_1 = -4$	$b'_4 = 1$	$a'_2 = 2(3\gamma^4 - \beta\gamma^2 + 3)$
	$b'_2 = 6$	$a'_3 = 2(2\gamma^4 - \alpha\gamma^3 + \alpha\gamma - 2)$
	$(c_n = c'_n / D)$	$a'_4 = \gamma^4 - \alpha\gamma^3 + \beta\gamma^2 - \alpha\gamma + 1$

Example: f_c = 500 Hz, F_s = 20 kHz. For this example, let's compute the fourth order case, so we can see how the various intermediate variables work. First is γ, which is computed just as we did above, although for the new set of input parameters: $\gamma = \tan\left(\pi f_c / F_s\right)$

$= \tan\left(\pi 500 / 20000\right) = \tan\left(\pi/40\right) = \gamma = 0.078701706824618$.

The values of α and β are included in the table, along with an exact formula for them should that be desired.

Now, with this information, we can compute D:
$D = \gamma^4 + \alpha\gamma^3 + \beta\gamma^2 + \alpha\gamma + 1$ = 0.078701706824618⁴ + 2.613125929752753 * 0.078701706824618³ + 3.414213562373095 * 0.078701706824618² + 2.613125929752753 * 0.078701706824618 + 1 = D = 1.228117167466711.

From there, we can compute the values for each coefficient:

$b_0 = b'_0 / D = 1/D = 0.814254556886247$; b_1 = -4 / D = -3.257018227544990

b_2 = 6 / D = 4.885527341317485; b_3 = -4 / D = b_1 = -3.257018227544990

b_4 = 1 / D = b_0 = 0.814254556886247.

The coefficients for the other orders are computed in a similar fashion and are shown in **Table 3** along with those just computed; frequency response plots follow in **Figure 74**.

1st-Ord	$b_0 = 0.927040342731733 \quad b_1 = -0.927040342731733$ $a_1 = -0.854080685463467$
2nd-Ord	$b_0 = b_2 = 0.894858606122573 \quad b_1 = -1.789717212245146$ $a_1 = -1.778631777824585 \quad a_2 = 0.800802646665708$
3rd-Ord	$b_0 = 0.854497231602542 \quad b_1 = -2.563491694807627$ $b_2 = 2.563491694807627 \quad b_3 = -0.854497231602542$ $a_1 = -2.686157396548144 \quad a_2 = 2.419655110966473$ $a_3 = -0.730165345305723$
4th-Ord	$b_0 = b_4 = 0.814254556886247 \quad b_2 = 4.885527341317480$ $b_1 = b_3 = -3.257018227544986$ $a_1 = -3.589733887112176 \quad a_2 = 4.851275882519418$ $a_3 = -2.924052656162460 \quad a_4 = 0.663010484385891$

Table 3 Coefficients: Butterworth Highpass Filters

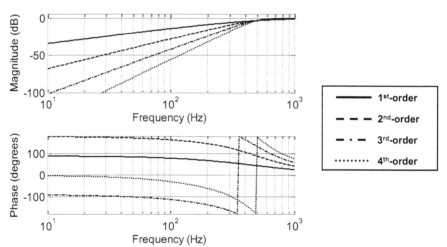

Figure 74 Frequency Responses of Butterworth Highpass Filters

2.3.1.3 Butterworth Bandpass Filters

The bandpass filters in this book have the same order cutoff on either side since the lowpass to bandpass transform doubles the filter order. In other words, the bandpass filter orders are even. We consider the second- and fourth-order Butterworth bandpass cases, which arise from the first- and second-order analog prototypes.

Second Order Butterworth Bandpass			
Definitions: $\gamma = \tan\left(\pi f_c / F_s\right)$		$D = (1+\gamma^2) f_c + \gamma BW$ $(c_n = c'_n / D)$	
Numerator Coefficients		Denominator Coefficients	
$b'_0 = BW\gamma$	$b_1 = 0$	$a'_1 = 2f_c(\gamma^2 - 1)$	$a'_2 = (1+\gamma^2) f_c - \gamma BW$
	$b'_2 = -b'_0$		

Note that $b'_1 = 0$, b_1 will also be 0, so b_1 is tabled; there is no need to divide (0) by D (though if you did, nothing would change!).

Fourth Order Butterworth Bandpass	
Definitions: $\gamma = \tan\left(\pi f_c / F_s\right)$ $(c_n = c'_n / D)$	
$D = f_c^2(\gamma^4 + 2\gamma^2 + 1) + \sqrt{2} BW f_c \gamma(\gamma^2 + 1) + BW^2 \gamma^2$	
Numerator Coefficients	Denominator Coefficients
$b'_0 = BW^2 \gamma^2$	$a'_1 = 2\left(2f_c^2(\gamma^4 - 1) + \sqrt{2} BW f_c \gamma(\gamma^2 - 1)\right)$
$b_1 = b_3 = 0$	$a'_2 = 2\left(3f_c^2(\gamma^4 + 1) - \gamma^2(2f_c^2 + BW^2)\right)$
$b'_2 = -2b'_0$	$a'_3 = 2\left(2f_c^2(\gamma^4 - 1) + \sqrt{2} BW f_c \gamma(1 - \gamma^2)\right)$
$b'_4 = b'_0$	$a'_4 = f_c^2(\gamma^4 + 2\gamma^2 + 1) - \sqrt{2} BW f_c \gamma(\gamma^2 + 1) + BW^2 \gamma^2$

Example: f_c = 5 kHz, BW = 10 kHz, F_s = 480 kHz. Recall from the **Theory** section that the Q of a filter is the ratio of its center frequency and bandwidth, $Q = f_c / BW$. The filters of this example have Qs of 0.5. The coefficients for this case are shown in **Table 4**; frequency response plots follow in **Figure 75**.

2nd-Ord	$b_0 = 0.061388151992198 \quad b_1 = 0 \quad b_2 = -0.061388151992198$
	$a_1 = -1.873204415984123 \quad a_2 = 0.877223696015605$
4th-Ord	$b_0 = b_4 = 0.003900146085567 \quad b_2 = -0.007800292171135$
	$b_1 = b_3 = 0$
	$a_1 = -3.807563717092115 \quad a_2 = 5.447200486220440$
	$a_3 = -3.470954129954859 \quad a_4 = 0.831334079749083$

Table 4 Coefficients: Butterworth Bandpass Filters

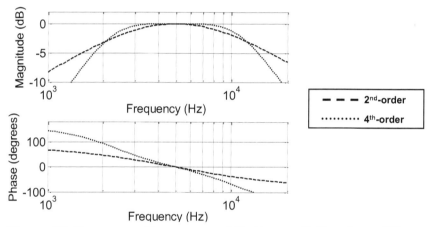

Figure 75 Frequency Responses of Butterworth Bandpass Filters

2.3.1.4 Butterworth Bandstop Filters

The bandstop filters in this book also have the same order cutoff on either side since the lowpass to bandstop transform also doubles the filter order. Therefore, as with the bandpass filters, we consider the second- and fourth-order cases, which arise from the first- and second-order analog prototypes.

Second Order Butterworth Bandstop	
Definitions: $\gamma = \tan\left(\pi f_c / F_s\right)$ $D = (1+\gamma^2)f_c + \gamma BW$ $(c_n = c'_n/D)$	
Numerator Coefficients	$b'_0 = f_c(\gamma^2+1)$ $b'_1 = 2f_c(\gamma^2-1)$ $b'_2 = b'_0$
Denominator Coefficients	$a'_1 = b'_1$ $a'_2 = (1+\gamma^2)f_c - \gamma BW$

Fourth Order Butterworth Bandstop

Definitions: $\gamma = \tan\left(\pi f_c / F_s\right)$ $(c_n = c'_n/D)$

$$D = f_c^2(\gamma^4 + 2\gamma^2 + 1) + \sqrt{2}BWf_c\gamma(\gamma^2+1) + BW^2\gamma^2$$

Numerator Coefficients

$b'_0 = f_c^2(\gamma^4 + 2\gamma^2 + 1)$ $b'_3 = b'_1$

$b'_1 = 4f_c^2(\gamma^4 - 1)$ $b'_4 = b'_0$

$$b'_2 = 2f_c^2(3\gamma^4 - 2\gamma^2 + 3)$$

Denominator Coefficients

$$a'_1 = 2\left(2f_c^2(\gamma^4 - 1) + \sqrt{2}BWf_c\gamma(\gamma^2 - 1)\right)$$

$$a'_2 = 2\left(3f_c^2(\gamma^4 + 1) - \gamma^2(2f_c^2 + BW^2)\right)$$

$$a'_3 = 2\left(2f_c^2(\gamma^4 - 1) + \sqrt{2}BWf_c\gamma(1 - \gamma^2)\right)$$

$$a'_4 = f_c^2(\gamma^4 + 2\gamma^2 + 1) - \sqrt{2}BWf_c\gamma(\gamma^2 + 1) + BW^2\gamma^2$$

Example: f_c = 100 kHz, BW = 20 kHz (Q = 5), F_s = 1 MHz. Coefficients are shown in **Table 5**; frequency response plots follow in **Figure 76**.

2nd-Ord	$b_0 = b_2 = 0.944484588770328 \quad b_1 = -1.528208166480858$ $a_1 = -1.528208166480858 \quad a_2 = 0.888969177540656$
4th-Ord	$b_0 = b_4 = 0.920318542147553 \quad b_2 = 4.250062308114151$ $b_1 = b_3 = -2.978213363342986$ $a_1 = -3.101995933368400 \quad a_2 = 4.243703063390296$ $a_3 = -2.854430793317572 \quad a_4 = 0.846996329018961$

Table 5 Coefficients: Butterworth Bandstop Filters

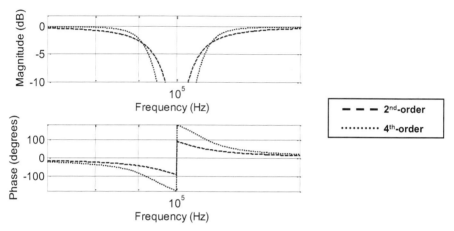

Figure 76 Frequency Responses of Butterworth Bandpass Filters

2.3.2 Linkwitz-Reilly Filters [Reference 17]

These filters are commonly used for audio crossover filters because a highpass and a lowpass with the same cutoff have magnitude responses that add to one so they keep the audio energy uniform across the spectrum.

Linkwitz-Reilly filters are formed by cascading two Butterworth filters of the same order. Therefore, Linkwitz-Reilly filters must have even order. For instance, the second-order Linkwitz-Reilly filter is constructed from the first-order Butterworth filter as follows: $\dfrac{1}{s+1} \cdot \dfrac{1}{s+1} = \dfrac{1}{s^2 + 2s + 1}$. The difference in the Butterworth and Linkwitz-Reilly second-order filters is that the latter has a $2s$ in the denominator in place of the $\sqrt{2}s$ found in the former.

In this section we provide simple design equations for the second- and fourth-order lowpass and highpass cases, as well as the fourth-order bandpass and bandstop filters.

2.3.2.1 Linkwitz-Reilly Lowpass Filters

Design equations for Linkwitz-Reilly lowpass filters are presented.

Second Order Linkwitz-Reilly Lowpass			
Definitions: $\gamma = \tan\left(\pi f_c / F_s\right)$		$D = \gamma^2 + 2\gamma + 1 \quad (c_n = c'_n / D)$	
Numerator Coefficients		Denominator Coefficients	
$b'_0 = \gamma^2$	$b'_1 = 2b'_0$	$b'_2 = b'_0$	$a'_1 = 2(\gamma^2 - 1)$	$a'_2 = \gamma^2 - 2\gamma + 1$

Fourth Order Linkwitz-Reilly Lowpass	
Definitions: $\gamma = \tan\left(\pi f_c / F_s\right)$	$D = \gamma^4 + 2\sqrt{2}\gamma^3 + 4\gamma^2 + 2\sqrt{2}\gamma + 1$
Numerator Coefficients	Denominator Coefficients
$b'_0 = \gamma^4 \qquad b'_3 = 4b'_0$	$a'_1 = 4\left(\gamma^4 + \sqrt{2}\gamma^3 - \sqrt{2}\gamma - 1\right)$
$b'_1 = 4b'_0 \qquad b'_4 = b'_0$	$a'_2 = 2\left(3\gamma^4 - 4\gamma^2 + 3\right)$
$b'_2 = 6b'_0$	$a'_3 = 4\left(\gamma^4 - \sqrt{2}\gamma^3 + \sqrt{2}\gamma - 1\right)$
$(c_n = c'_n / D)$	$a'_4 = \gamma^4 - 2\sqrt{2}\gamma^3 + 4\gamma^2 - 2\sqrt{2}\gamma + 1$

Since a Linkwitz-Reilly filter's likely use is as an audio crossover, let us try an example of lowpass filters with cutoffs of 100 Hz, each operating at a sample rate of 48 kHz. Coefficients are shown in **Table 6**; plots follow in **Figure 77**.

2nd-Ord	$b_0 = b_2 = 0.422827489135607 \times 10^{-4} \quad b_1 = 0.845654978271213 \times 10^{-4}$ $a_1 = -1.973989925363103 \quad a_2 = 0.974159056358757$
4th-Ord	$b_0 = b_4 = 0.018014394990860 \times 10^{-7} \quad b_2 = 0.108086369945159 \times 10^{-7}$ $b_1 = b_3 = 0.072057579963440 \times 10^{-7}$ $a_1 = -3.962977018289147 \quad a_2 = 5.889613277106252$ $a_3 = -3.890289213824895 \quad a_4 = 0.963652983830821$

Table 6 Coefficients: Linkwitz-Reilly Lowpass Filters

Although difficult to read precisely from this plot, Linkwitz-Reilly filters have magnitudes of -6 dB at their cutoff frequencies, unlike the -3 dB cutoffs of many filters.

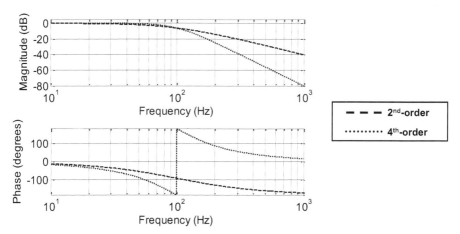

Figure 77 Frequency Responses of Linkwitz-Reilly Lowpass Filters

2.3.2.2 Linkwitz-Reilly Highpass Filters

Design equations for Linkwitz-Reilly highpass filters follow:

Second Order Linkwitz-Reilly Highpass				
Definitions: $\gamma = \tan\left(\pi f_c / F_s\right)$			$D = \gamma^2 + 2\gamma + 1$	$(c_n = c'_n/D)$
Numerator Coefficients			Denominator Coefficients	
$b'_0 = 1$	$b'_1 = -2$	$b'_2 = 1$	$a'_1 = 2(\gamma^2 - 1)$	$a'_2 = \gamma^2 - 2\gamma + 1$
Fourth Order Linkwitz-Reilly Highpass				
Definitions: $\gamma = \tan\left(\pi f_c / F_s\right)$			$D = \gamma^4 + 2\sqrt{2}\gamma^3 + 4\gamma^2 + 2\sqrt{2}\gamma + 1$	
Numerator Coefficients			Denominator Coefficients	
$b'_0 = 1$		$b'_3 = -4$	$a'_1 = 4\left(\gamma^4 + \sqrt{2}\gamma^3 - \sqrt{2}\gamma - 1\right)$	
$b'_1 = -4$		$b'_4 = 1$	$a'_2 = 2\left(3\gamma^4 - 4\gamma^2 + 3\right)$	
$b'_2 = 6$			$a'_3 = 4\left(\gamma^4 - \sqrt{2}\gamma^3 + \sqrt{2}\gamma - 1\right)$	
$(c_n = c'_n/D)$			$a'_4 = \gamma^4 - 2\sqrt{2}\gamma^3 + 4\gamma^2 - 2\sqrt{2}\gamma + 1$	

Chapter 3 IIR Filters

To form crossover pairs with the lowpass filter examples designed above, let us try highpass filters with cutoffs of 100 Hz, each operating at a sample rate of 48 kHz. Coefficients are shown in **Table 7**; frequency response plots follow in **Figure 78**.

2nd-Ord	$b_0 = b_2 = 0.987037245430465 \quad b_1 = -1.974074490860930$ $a_1 = -1.973989925363103 \quad a_2 = 0.974159056358757$
4th-Ord	$b_0 = b_4 = 0.981658280815695 \quad b_2 = 5.889949684894169$ $b_1 = b_3 = -3.926633123262779$ $a_1 = -3.962977018289147 \quad a_2 = 5.889613277106252$ $a_3 = -3.890289213824895 \quad a_4 = 0.963652983830821$

Table 7 Coefficients: Linkwitz-Reilly Highpass Filters

The denominator coefficients are the same as for the lowpass case. This is often true for lowpass-highpass pairs.

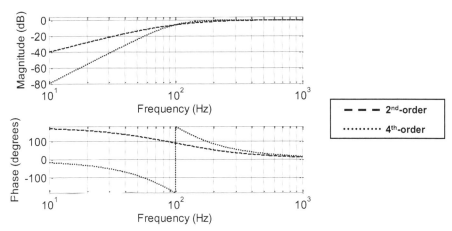

Figure 78 Frequency Responses of Linkwitz-Reilly Highpass Filters

2.3.2.3 Linkwitz-Reilly Bandpass Filters

The lowest order Linkwitz-Reilly analog lowpass filter is second-order, and the lowpass to bandpass transform doubles the order. Therefore, the lowest order bandpass is fourth-order, as shown here.

Fourth Order Linkwitz-Reilly Bandpass	
Definitions:	
$\gamma = \tan\left(\pi f_c / F_s\right)$ $D = f_c^2\left(\gamma^4 + 2\gamma^2 + 1\right) + 2BWf_c\gamma\left(\gamma^2 + 1\right) + BW^2\gamma^2$	
Numerator Coefficients	Denominator Coefficients
$b_0' = BW^2\gamma^2$	$a_1' = 4\left(f_c^2\left(\gamma^4 - 1\right) + BWf_c\gamma\left(\gamma^2 - 1\right)\right)$
$b_1 = 0$ $b_3 = 0$	$a_2' = 2\left(3f_c^2\left(\gamma^4 + 1\right) - \gamma^2\left(2f_c^2 + BW^2\right)\right)$
$b_2' = -2b_0'$ $b_4' = b_0'$	$a_3' = 4\left(f_c^2\left(\gamma^4 - 1\right) + BWf_c\gamma\left(1 - \gamma^2\right)\right)$
$\left(c_n = c_n'/D\right)$	$a_4' = f_c^2\left(\gamma^4 + 2\gamma^2 + 1\right) - 2BWf_c\gamma\left(\gamma^2 + 1\right) + BW^2\gamma^2$

Example: f_c = 1 kHz, BW = 2 kHz (Q = 0.5), F_s = 192 kHz. Coefficients are shown in **Table 8**; frequency response plots follow in **Figure 79**.

4th-Ord	$b_0 = b_4 = 0.001003778381422$ $b_2 = -0.002007556762845$ $b_1 = b_3 = 0$ $a_1 = -3.871196355723198$ $a_2 = 5.619810459211712$ $a_3 = -3.625898294925914$ $a_4 = 0.877285266596260$

Table 8 Coefficients for Linkwitz-Reilly Bandpass Filters

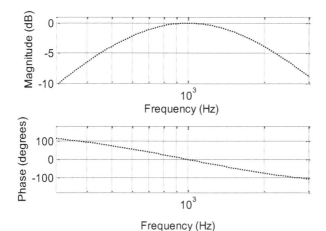

Figure 79 Frequency Responses of Linkwitz-Reilly Bandpass Filter

2.3.2.4 Linkwitz-Reilly Bandstop Filters

For the same reasons as above, the lowest order Linkwitz-Reilly order bandstop filter is fourth order; the design equations for this filter are shown here.

Fourth Order Linkwitz-Reilly Bandstop
Definitions: $\gamma = \tan\left(\pi f_c / F_s\right)$ $\quad (c_n = c'_n / D)$
$D = f_c^2\left(\gamma^4 + 2\gamma^2 + 1\right) + 2BWf_c\gamma\left(\gamma^2 + 1\right) + BW^2\gamma^2$
Numerator Coefficients
$b'_0 = f_c^2\left(\gamma^4 + 2\gamma^2 + 1\right)$ $\qquad\qquad b'_3 = b'_1$
$b'_1 = 4f_c^2\left(\gamma^4 - 1\right)$ $\qquad\qquad\qquad b'_4 = b'_0$
$b'_2 = 2f_c^2\left(3\gamma^4 - 2\gamma^2 + 3\right)$
Denominator Coefficients
$a'_1 = 4\left(f_c^2\left(\gamma^4 - 1\right) + BWf_c\gamma\left(\gamma^2 - 1\right)\right)$
$a'_2 = 2\left(3f_c^2\left(\gamma^4 + 1\right) - \gamma^2\left(2f_c^2 + BW^2\right)\right)$
$a'_3 = 4\left(f_c^2\left(\gamma^4 - 1\right) + BWf_c\gamma\left(1 - \gamma^2\right)\right)$
$a'_4 = f_c^2\left(\gamma^4 + 2\gamma^2 + 1\right) - 2BWf_c\gamma\left(\gamma^2 + 1\right) + BW^2\gamma^2$

Example: f_c = 10 kHz, BW = 0.5 kHz (Q = 20), F_s = 176 kHz. Coefficients are shown in **Table 9**; frequency response plots follow in **Figure 80**.

4th-Ord	$b_0 = b_4 = 0.982753137222581 \quad b_2 = 5.416443079192880$ $b_1 = b_3 = -3.683161126653403$ $a_1 = -3.715339448689833 \quad a_2 = 5.416293055522837$ $a_3 = -3.650982804616972 \quad a_4 = 0.965656298115207$

Table 9 Coefficients for Linkwitz-Reilly Bandstop Filters

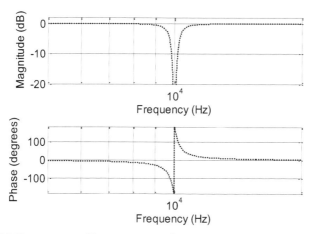

Figure 80 Frequency Responses of Linkwitz-Reilly Bandstop Filter

2.3.3 Bessel Filters [Reference 5]

Bessel filters are known for their passband phase linearity, leading to very nearly constant group delays. While there are also advanced methods of designing linear phase IIR filters, as discussed in the **Advanced Topics** section, Bessel filters are perhaps the most approachable IIR method for phase linearity.

Bessel analog filters are designed using the following equation

$$H(s) = \frac{B_0}{\sum_{i=0}^{n} B_i s^i} \qquad B_i = \frac{(2n-i)!}{2^{n-i} i!(n-i)!}$$

This equation leads to the analog prototypes from which we can design the various digital Bessel filters using the process described in the **Theory** section of this book.

First-order Bessel filters turn out to be equivalent to first-order Butterworth filters, so only the second- and higher-order Bessel filters are shown in this section.

2.3.3.1 Bessel Lowpass Filters

Design equations for second- through fourth-order Bessel lowpass filters are shown below.

Second Order Bessel Lowpass

Definitions: $\gamma = \tan\left(\pi f_c / F_s\right)$ $D = 3\gamma^2 + 3\gamma + 1$ $(c_n = c'_n/D)$

Numerator Coefficients			Denominator Coefficients	
$b'_0 = 3\gamma^2$	$b'_1 = 2b'_0$	$b'_2 = b'_0$	$a'_1 = 2(3\gamma^2 - 1)$	$a'_2 = 3\gamma^2 - 3\gamma + 1$

Third Order Bessel Lowpass

Definitions: $\gamma = \tan\left(\pi f_c / F_s\right)$ $D = 15\gamma^3 + 15\gamma^2 + 6\gamma + 1$ $(c_n = c'_n/D)$

Numerator Coefficients		Denominator Coefficients
$b'_0 = 15\gamma^3$	$b'_2 = 3b'_0$	$a'_1 = 3(15\gamma^3 + 5\gamma^2 - 2\gamma - 1)$
$b'_1 = 3b'_0$	$b'_3 = b'_0$	$a'_2 = 3(15\gamma^3 - 5\gamma^2 - 2\gamma + 1)$
$(c_n = c'_n/D)$		$a'_3 = 15\gamma^3 - 15\gamma^2 + 6\gamma - 1$

Fourth Order Bessel Lowpass

Definitions: $\gamma = \tan\left(\pi f_c / F_s\right)$ $D = 105\gamma^4 + 105\gamma^3 + 45\gamma^2 + 10\gamma + 1$

Numerator Coefficients		Denominator Coefficients
$b'_0 = 105\gamma^4$	$b'_3 = 4b'_0$	$a'_1 = 2(210\gamma^4 + 105\gamma^3 - 10\gamma - 2)$
$b'_1 = 4b'_0$	$b'_4 = b'_0$	$a'_2 = 6(105\gamma^4 - 15\gamma^2 + 1)$
$b'_2 = 6b'_0$		$a'_3 = 2(210\gamma^4 - 105\gamma^3 + 10\gamma - 2)$
$(c_n = c'_n/D)$		$a'_4 = 105\gamma^4 - 105\gamma^3 + 45\gamma^2 - 10\gamma + 1$

Example: f_c = 100 kHz, F_s = 2 MHz. Coefficients are shown in **Table 10**; frequency response plots follow in **Figure 81**. The phase plot in **Figure 81** is a linear plot across the passband. The magnitude plot shows the transition region, which is of the most interest for this plot. Note that this filter does not have a -3 dB cutoff; the cutoff and the group delay can be traded off as desired by slightly altering f_c.

As discussed in the **Theory** section, phase linearity leads to constant group delays. The group delay from 0 Hz (DC) to the Nyquist frequency is shown on the left in **Figure 82**. (The line styles correspond to the previous plot.) Clearly, this is not a constant group delay. However, on the right is the passband of the filter only, which corresponds to the portion of the phase plotted in **Figure 81**. This is quite constant, showing only a fraction of a sample deviation across the entire passband.

2nd-Ord	$b_0 = b_2 = 0.048539987766887$ $\quad b_1 = 0.097079975533774$ $a_1 = -1.192901206361467$ $\quad a_2 = 0.387061157429015$
3rd-Ord	$b_0 = b_3 = 0.024976064963405$ $\quad b_1 = b_2 = 0.074928194890215$ $a_1 = -1.422867081129509$ $\quad a_2 = 0.776217613793158$ $a_3 = -0.153542012956410$
4th-Ord	$b_0 = b_4 = 0.015747383127697$ $\quad b_2 = 0.094484298766184$ $b_1 = b_3 = 0.062989532510789$ $a_1 = -1.446397555336502$ $\quad a_2 = 0.986365170938403$ $a_3 = -0.334220903183651$ $\quad a_4 = 0.046211417624906$

Table 10 Coefficients for Bessel Lowpass Filters: Various Orders

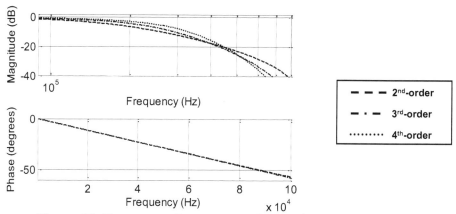

Figure 81 Frequency Responses of Bessel Lowpass Filters

Sometimes it is desirable to compensate for a filter's delay, which is difficult to do when that is not constant. Even when it is constant, if it is not an integer number of samples, such as in our example, a simple delay will not suffice. In such a case, the cutoff frequency of the Bessel filter can be modified slightly to produce an integer delay. For instance, rather than the cutoff of 100 kHz of our example, a cutoff of 105.15 kHz produces a delay that is very nearly 3 samples.

Chapter 3 IIR Filters

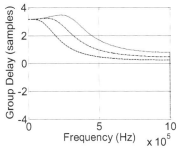
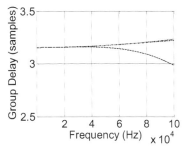

Figure 82 Group Delays of Bessel Lowpass Filters

2.3.3.2 Bessel Highpass Filters

Note that the Bessel is the first filter we have seen where the lowpass filter denominators are different than for the highpass:

Second Order Bessel Highpass				
Definitions: $\gamma = \tan\left(\pi f_c / F_s\right)$			$D = \gamma^2 + 3\gamma + 3$	$(c_n = c'_n / D)$
Numerator Coefficients			Denominator Coefficients	
$b'_0 = 3$	$b'_1 = -6$	$b'_2 = 3$	$a'_1 = 2(\gamma^2 - 3)$	$a'_2 = \gamma^2 - 3\gamma + 3$
Third Order Bessel Highpass			
Definitions: $\gamma = \tan\left(\pi f_c / F_s\right)$		$D = \gamma^3 + 6\gamma^2 + 15\gamma + 15$	$(c_n = c'_n / D)$
Numerator Coefficients		Denominator Coefficients	
$b'_0 = 15$	$b'_2 = 45$	$a'_1 = 3(\gamma^3 + 2\gamma^2 - 5\gamma - 15)$	
$b'_1 = -45$	$b'_3 = -15$	$a'_2 = 3(\gamma^3 - 2\gamma^2 - 5\gamma + 15)$	
		$a'_3 = \gamma^3 - 6\gamma^2 + 15\gamma - 15$	
Fourth Order Bessel Highpass			
Definition: $\gamma = \tan\left(\pi f_c / F_s\right)$		$D = \gamma^4 + 10\gamma^3 + 45\gamma^2 + 105\gamma + 105$	
Numerator Coefficients		Denominator Coefficients	
$b'_0 = 105$	$b'_3 = -420$	$a'_1 = 2(2\gamma^4 + 10\gamma^3 - 105\gamma - 210)$	
$b'_1 = -420$	$b'_4 = 105$	$a'_2 = 6(\gamma^4 - 15\gamma^2 + 105)$	
$b'_2 = 630$		$a'_3 = 2(2\gamma^4 - 10\gamma^3 + 105\gamma - 210)$	
$(c_n = c'_n / D)$		$a'_4 = \gamma^4 - 10\gamma^3 + 45\gamma^2 - 105\gamma + 105$	

Example: f_c = 3.5 kHz, F_s = 14 kHz. Coefficients are shown in **Table 11**. Note that for the third-order filter the b_0 and b_3 and b_1 and b_2 signs are opposite. Frequency response plots follow in **Figure 83**. Note that the phase plot in **Figure 83** is for the passband region plotted on a linear frequency axis; the response is not quite as linear as for the lowpass case above. The magnitude plot shows the transition region, which is of the most interest for this plot.

2nd-Ord	$b_0 = b_2 = 0.428571428571429 \quad b_1 = -0.857142857142857$
	$a_1 = -0.571428571428572 \quad a_2 = 0.142857142857143$
3rd-Ord	$b_0 = -b_3 = 0.405405405405405 \quad -b_1 = b_2 = 1.216216216216216$
	$a_1 = -1.378378378378378 \quad a_2 = 0.729729729729730$
	$a_3 = -0.135135135135135$
4th-Ord	$b_0 = b_4 = 0.394736842105263 \quad b_2 = 2.368421052631579$
	$b_1 = b_3 = -1.578947368421053$
	$a_1 = -2.278195488721805 \quad a_2 = 2.052631578947369$
	$a_3 = -0.849624060150376 \quad a_4 = 0.135338345864662$

Table 11 Coefficients for Bessel Highpass Filters: Various Orders

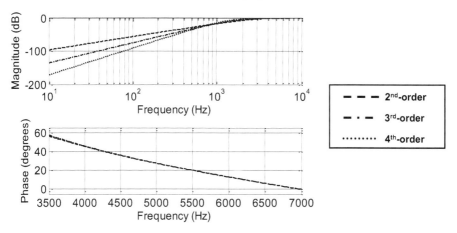

Figure 83 Frequency Responses of Bessel Highpass Filters

While Bessel filters have some of the flattest group delays available from standard IIR filters, they are by no means perfect, as seen in **Figure 84** below. (Line styles correspond to those of the previous plot.) And they can be even worse for some cases. In the **Advanced**

Topics section, we do discuss design techniques for IIR filters with phases that are even more linear than for Bessel filters, though Bessel filters are completely adequate for many applications.

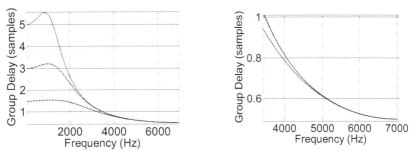

Figure 84 Group Delays of Bessel Highpass Filters:
Full Band (L), Passband (R)

2.3.3.3 Bessel Bandpass Filters

Since the first-order Bessel is the same as the first-order Butterworth the second-order bandpasses are equivalent. Shown is the fourth-order bandpass, arising from the second-order Bessel lowpass.

Fourth Order Bessel Bandpass				
Definitions: $\gamma = \tan\left(\pi f_c / F_s\right)$ $\quad (c_n = c'_n/D)$				
$D = f_c^2 \gamma^4 + 3BWf_c\gamma^3 + \left(2f_c^2 + 3BW^2\right)\gamma^2 + 3BWf_c\gamma + f_c^2$				
Numerator Coefficients				
$b'_0 = 3BW^2\gamma^2$	$b'_1 = 0$	$b'_2 = -2b'_0$	$b'_3 = 0$	$b'_4 = b'_0$
Denominator Coefficients				
$a'_1 = 2f_c\left(2f_c\gamma^4 + 3BW\left(\gamma^3 - \gamma\right) - 2f_c\right)$				
$a'_2 = 2\left(3f_c^2\gamma^4 - \left(2f_c^2 + 3BW^2\right)\gamma^2 + 3f_c^2\right)$				
$a'_3 = 2f_c\left(2f_c\gamma^4 - 3BW\left(\gamma^3 - \gamma\right) - 2f_c\right)$				
$a'_4 = f_c^2\gamma^4 - 3BWf_c\gamma^3 + \left(2f_c^2 + 3BW^2\right)\gamma^2 - 3BWf_c\gamma + f_c^2$				

Example: f_c = 500 Hz, BW = 600 Hz (Q=5/6), F_s = 2.5 kHz. Coefficients for this case are shown in **Table 12**; a frequency response plot follows in **Figure 85**. Note that the phase plot in

Figure 85 is a linear frequency plot across the passband, showing the phase linearity. The magnitude plot shows the transition region. The group delay for this filter is shown in **Figure 86**.

Note also that, while this group delay is not constant, it is very well behaved, particularly in the passband, shown at right.

4th- Ord	$b_0 = b_4 = 0.264822403390719 \quad b_2 = -0.529644806781438$	
	$b_1 = b_3 = 0$	
	$a_1 = -0.621909435392721 \quad a_2 = 0.116089523639032$	
	$a_3 = -0.048269336290184 \quad a_4 = 0.071830822342236$	

Table 12 Coefficients for Bessel Bandpass Filter

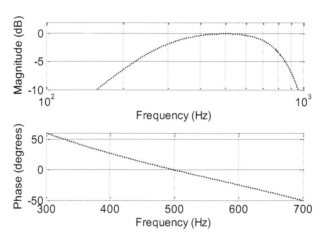

Figure 85 Frequency Response of a Bessel Bandpass Filter

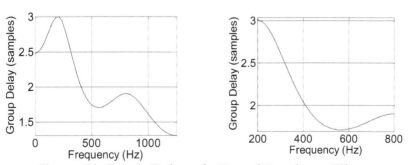

Figure 86 Group Delay of a Bessel Bandpass Filter: Full Band (L), Passband (R)

2.3.3.4 Bessel Bandstop Filters

Since the first-order Bessel is the same as the first-order Butterworth, the reader is referred to that section for the second-order bandstop, which arises from a first-order lowpass. The fourth-order bandstop, arising from the second-order Bessel lowpass, is show below.

Fourth Order Bessel Bandstop		
Definitions: $\gamma = \tan\left(\pi f_c / F_s\right)$ $\quad (c_n = c'_n / D)$		
$D = 3f_c^2\gamma^4 + 3BWf_c\gamma^3 + \left(6f_c^2 + BW^2\right)\gamma^2 + 3BWf_c\gamma + 3f_c^2$		
Numerator Coefficients		
$b'_0 = 3f_c^2\left(\gamma^4 + 2\gamma^2 + 1\right)$	$b'_1 = 12f_c^2\left(\gamma^4 - 1\right)$	
$b'_2 = 6f_c^2\left(3\gamma^4 - 2\gamma^2 + 3\right)$	$b'_3 = b'_1$	$b'_4 = b'_0$
Denominator Coefficients		
$a'_1 = 6f_c\left(2f_c\gamma^4 + BW\left(\gamma^3 - \gamma\right) - 2f_c\right)$		
$a'_2 = 2\left(9f_c^2\gamma^4 - \left(6f_c^2 + BW^2\right)\gamma^2 + 9f_c^2\right)$		
$a'_3 = 6f_c\left(2f_c\gamma^4 - BW\left(\gamma^3 - \gamma\right) - 2f_c\right)$		
$a'_4 = 3f_c^2\gamma^4 - 3BWf_c\gamma^3 + \left(6f_c^2 + BW^2\right)\gamma^2 - 3BWf_c\gamma + 3f_c^2$		

Example: f_c = 3 kHz, BW = 500 Hz (Q = 6), F_s = 12.5 kHz. Coefficients are shown in **Table 13**; the frequency response plot follows in **Figure 87**, which shows the magnitude response in the transition region (log frequency) and the passband phase linearity (linear frequency).

4th-Ord	$b_0 = b_4 = 0.921256025947146 \quad b_2 = 1.857040809755153$
	$b_1 = b_3 = -0.231384577954928$
	$a_1 = -0.241006577677641 \quad a_2 = 1.852792551252922$
	$a_3 = -0.221762578232215 \quad a_4 = 0.846760310396523$

Table 13 Coefficients for Bessel Bandstop Filter

The group delay for this filter is shown in **Figure 88**. Note that while this is not constant, it is very well behaved. Notice that outside the stopband, the group delay is generally a fraction of a sample.

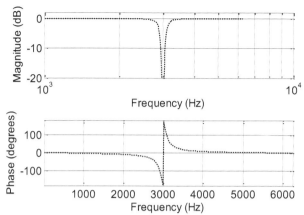

Figure 87 Frequency Response of a Bessel Bandstop Filter

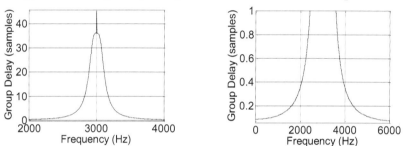

Figure 88 Group Delay of a Bessel Bandstop Filter: Full Band (L), Passband (R)

2.3.4 Chebychev Type I Filters

Unlike Butterworth, Linkwitz-Reilly and Bessel filters, Chebychev filters are allowed to have significant ripple either in the passband (Type I) or in the stopband (Type II). These filters are characterized by quick transitions but also by increased phase distortion. You might find their unique shapes, and particularly their quick transitions, useful for some applications.

Besides cutoff frequency and sample rate, it is necessary in the design of Chebychev filters to specify the maximum allowable ripple, R. To design these filters we manipulate R to determine v_0, and from that intermediate variables κ and λ that are used in the filter coefficient definitions. Note that v_0, κ and λ depend upon filter order; they are not the same for each of the orders considered.

2.3.4.1 Type I Chebychev Lowpass Filters

Here are design equations for Type I Chebychev lowpass filters.

First Order Chebychev Type I Lowpass

Definitions: $(c_n = c'_n / D)$

$$v_0 = \operatorname{arcsinh}\left(\frac{1}{\sqrt{10^{\frac{R}{10}} - 1}}\right) \quad \kappa = \sinh(v_0) \quad \gamma = \tan\left(\pi f_c / F_s\right) \quad D = \kappa\gamma + 1$$

Numerator Coefficients		Denominator Coefficient
$b'_0 = \kappa\gamma$	$b'_1 = b'_0$	$a'_1 = \kappa\gamma - 1$

Second Order Chebychev Type I Lowpass

Definitions: $\quad v_0 = \dfrac{1}{2}\operatorname{arcsinh}\left(\dfrac{1}{\sqrt{10^{\frac{R}{10}} - 1}}\right) \quad \gamma = \tan\left(\pi f_c / F_s\right)$

$$\kappa = \sinh(v_0) \quad \lambda = \cosh(v_0) \quad D = (\kappa^2 + \lambda^2)\gamma^2 + 2\sqrt{2}\kappa\gamma + 2$$

Numerator Coefficients		Denominator Coefficients
$b'_0 = (\kappa^2 + \lambda^2)\gamma^2$		$a'_1 = 2(\kappa^2 + \lambda^2)\gamma^2 - 4$
$b'_1 = 2b'_0$	$b'_2 = b'_0$	$a'_2 = (\kappa^2 + \lambda^2)\gamma^2 - 2\sqrt{2}\kappa\gamma + 2$

Third Order Chebychev Type I Lowpass

Definitions: $(c_n = c'_n / D)$

$$v_0 = \frac{1}{3}\operatorname{arcsinh}\left(\frac{1}{\sqrt{10^{\frac{R}{10}} - 1}}\right) \quad \kappa = \sinh(v_0) \quad \lambda = \cosh(v_0)$$

$$\gamma = \tan\left(\pi f_c / F_s\right) \quad D = \kappa(\kappa^2 + 3\lambda^2)\gamma^3 + (5\kappa^2 + 3\lambda^2)\gamma^2 + 8\kappa\gamma + 4$$

Numerator Coefficients			
$b'_0 = \kappa(\kappa^2 + 3\lambda^2)\gamma^3$	$b'_1 = 3b'_0$	$b'_2 = 3b'_0$	$b'_3 = b'_0$

Denominator Coefficients
$a'_1 = 3\kappa(\kappa^2 + 3\lambda^2)\gamma^3 + (5\kappa^2 + 3\lambda^2)\gamma^2 - 8\kappa\gamma - 12$
$a'_2 = 3\kappa(\kappa^2 + 3\lambda^2)\gamma^3 - (5\kappa^2 + 3\lambda^2)\gamma^2 - 8\kappa\gamma + 12$
$a'_3 = \kappa(\kappa^2 + 3\lambda^2)\gamma^3 - (5\kappa^2 + 3\lambda^2)\gamma^2 + 8\kappa\gamma - 4$

Fourth Order Chebychev Type I Lowpass
Definitions: $(c_n = c'_n / D)$
$v_0 = \dfrac{1}{4}\mathrm{arcsinh}\left(\dfrac{1}{\sqrt{10^{\frac{R}{10}}-1}}\right)\quad \kappa = \sinh(v_0)\quad \lambda = \cosh(v_0)$
$\alpha = \cos\left(\dfrac{5\pi}{8}\right)\quad \beta = \cos\left(\dfrac{7\pi}{8}\right)\quad \gamma = \tan\left(\pi f_c / F_s\right)$
$D = \left(\dfrac{\kappa^4+\lambda^4}{8} + \dfrac{3\kappa^2\lambda^2}{4}\right)\gamma^4 - \kappa\left(\dfrac{\sqrt{2}}{2}\kappa^2(\alpha+\beta) + 2\lambda^2(\alpha^3+\beta^3)\right)\gamma^3$ $+\left(\left(1+\sqrt{2}\right)\kappa^2 + \lambda^2\right)\gamma^2 - 2\kappa(\alpha+\beta)\gamma + 1$
Numerator Coefficients

$b'_0 = \left(\dfrac{\kappa^4+\lambda^4}{8} + \dfrac{3\kappa^2\lambda^2}{4}\right)\gamma^4$	$b'_2 = 6\left(\dfrac{\kappa^4+\lambda^4}{8} + \dfrac{3\kappa^2\lambda^2}{4}\right)\gamma^4$
$b'_1 = 4\left(\dfrac{\kappa^4+\lambda^4}{8} + \dfrac{3\kappa^2\lambda^2}{4}\right)\gamma^4$	$b'_3 = 4\left(\dfrac{\kappa^4+\lambda^4}{8} + \dfrac{3\kappa^2\lambda^2}{4}\right)\gamma^4$

$b'_4 = \left(\dfrac{\kappa^4+\lambda^4}{8} + \dfrac{3\kappa^2\lambda^2}{4}\right)\gamma^4$
Denominator Coefficients
$a'_1 = \left(\dfrac{\kappa^4+\lambda^4}{2} + 3\kappa^2\lambda^2\right)\gamma^4 - \kappa\left(\sqrt{2}\kappa^2(\alpha+\beta) + 4\lambda^2(\alpha^3+\beta^3)\right)\gamma^3 + 4\kappa(\alpha+\beta)\gamma - 4$
$a'_2 = \left(3\dfrac{\kappa^4+\lambda^4}{4} + \dfrac{9\kappa^2\lambda^2}{2}\right)\gamma^4 - 2\left(\left(1+\sqrt{2}\right)\kappa^2 + \lambda^2\right)\gamma^2 + 6$
$a'_3 = \left(\dfrac{\kappa^4+\lambda^4}{2} + 3\kappa^2\lambda^2\right)\gamma^4 + \kappa\left(\sqrt{2}\kappa^2(\alpha+\beta) + 4\lambda^2(\alpha^3+\beta^3)\right)\gamma^3 - 4\kappa(\alpha+\beta)\gamma - 4$
$a'_4 = \left(\dfrac{\kappa^4+\lambda^4}{8} + \dfrac{3\kappa^2\lambda^2}{4}\right)\gamma^4 + \kappa\left(\dfrac{\sqrt{2}}{2}\kappa^2(\alpha+\beta) + 2\lambda^2(\alpha^3+\beta^3)\right)\gamma^3 +$ $\left(\left(1+\sqrt{2}\right)\kappa^2 + \lambda^2\right)\gamma^2 + 2\kappa(\alpha+\beta)\gamma + 1$

Example: f_c = 1550 Hz, F_s = 75 kHz, R = 1 dB ripple. Coefficients are shown in **Table 14**; frequency response plots follow in **Figure 89**.

Chapter 3 IIR Filters

1st-Ord	$b_0 = b_1 = 0.113297828670419$ $a_1 = -0.773404342659162$
2nd-Ord	$b_0 = b_2 = 0.004331312250899 \quad b_1 = 0.008662624501797$ $a_1 = -1.850016912692086 \quad a_2 = 0.867342161695681$
3rd-Ord	$b_0 = b_3 = 0.126244553794310 \times 10^{-3} \quad b_1 = b_2 = 0.378733661382930 \times 10^{-3}$ $a_1 = -2.859512236422960 \quad a_2 = 2.740116715784487$ $a_3 = -0.879594522931172$
4th-Ord	$b_0 = b_4 = 0.046105435736307 \times 10^{-4} \quad b_2 = 0.276632614417842 \times 10^{-4}$ $b_1 = b_3 = 0.184421742945228 \times 10^{-4}$ $a_1 = -3.859826707384402 \quad a_2 = 5.604896247200569$ $a_3 = -3.628636373139489 \quad a_4 = 0.883640602020501$

Table 14 Coefficients: T1 Chebychev Lowpass Filters

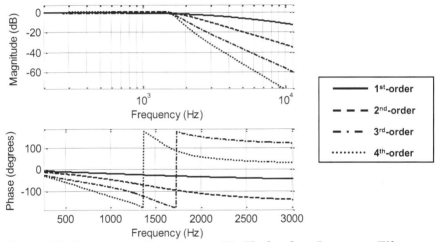

Figure 89 Frequency Responses: T1 Chebychev Lowpass Filters

Taking a closer look at the passband ripple, **Figure 90**, we see that the second- and fourth-order filters hit peaks of 1 dB above 0 dB, and hit the cutoff frequency at 0 dB.

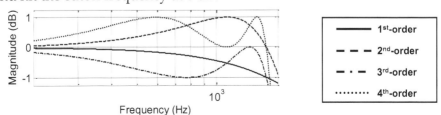

Figure 90 Passband Ripple of Type I Chebychev Lowpass Filters

Similarly, the first- and third-order filters hit passband minima of 1 dB below 0, and are at that point (again) at the cutoff frequency. All of these filters are designed to be exactly 0 dB at DC (0 Hz).

In some cases, it may be desirable to set the passband gain to exactly 0 dB at a critical point, or to set the gain at the cutoff frequency to some precise value. This is accomplished for any filter by scaling the numerator coefficients (only) by a (linear) gain adjustment.

2.3.4.2 Type I Chebychev Highpass Filters

Equations for Type I Chebychev highpass filters are presented.

First Order Chebychev Type I Highpass

Definitions: $v_0 = \operatorname{arcsinh}\left(\dfrac{1}{\sqrt{10^{\frac{R}{10}}-1}}\right)$ $\quad \kappa = \sinh(v_0) \quad D = \gamma + \kappa$

$\gamma = \tan\left(\pi f_c / F_s\right) \quad (c_n = c'_n / D)$

Numerator Coefficients		Denominator Coefficient
$b'_0 = \kappa$	$b'_1 = -b'_0$	$a'_1 = \gamma - \kappa$

Second Order Chebychev Type I Highpass

Definitions:

$v_0 = \dfrac{1}{2}\operatorname{arcsinh}\left(\dfrac{1}{\sqrt{10^{\frac{R}{10}}-1}}\right) \quad \kappa = \sinh(v_0) \quad \lambda = \cosh(v_0)$

$\gamma = \tan\left(\pi f_c / F_s\right) \quad D = 2\gamma^2 + 2\sqrt{2}\kappa\gamma + \kappa^2 + \lambda^2 \quad (c_n = c'_n / D)$

Numerator Coefficients		Denominator Coefficients
$b'_0 = \kappa^2 + \lambda^2$		$a'_1 = 4\gamma^2 - 2(\kappa^2 + \lambda^2)$
$b'_1 = -2b'_0$	$b'_2 = b'_0$	$a'_2 = 2\gamma^2 - 2\sqrt{2}\kappa\gamma + \kappa^2 + \lambda^2$

Third Order Chebychev Type I Highpass

Definitions: $\left(c_n = c'_n/D\right)$

$$v_0 = \frac{1}{3}\operatorname{arcsinh}\left(\frac{1}{\sqrt{10^{\frac{R}{10}}-1}}\right) \quad \kappa = \sinh(v_0) \quad \lambda = \cosh(v_0)$$

$$\gamma = \tan\left(\pi f_c / F_s\right) \quad D = 4\gamma^3 + 8\kappa\gamma^2 + \left(5\kappa^2 + 3\lambda^2\right)\gamma + \kappa\left(\kappa^2 + 3\lambda^2\right)$$

Numerator Coefficients

$b'_0 = \kappa\left(\kappa^2 + 3\lambda^2\right)$	$b'_2 = 3b'_0$
$b'_1 = -3b'_0$	$b'_3 = -b'_0$

Denominator Coefficients

$$a'_1 = 12\gamma^3 + 8\kappa\gamma^2 - \left(5\kappa^2 + 3\lambda^2\right)\gamma - 3\kappa\left(\kappa^2 + 3\lambda^2\right)$$

$$a'_2 = 12\gamma^3 - 8\kappa\gamma^2 - \left(5\kappa^2 + 3\lambda^2\right)\gamma + 3\kappa\left(\kappa^2 + 3\lambda^2\right)$$

$$a'_3 = 4\gamma^3 - 8\kappa\gamma^2 + \left(5\kappa^2 + 3\lambda^2\right)\gamma - \kappa\left(\kappa^2 + 3\lambda^2\right)$$

Fourth Order Chebychev Type I Highpass

Definitions: $\left(c_n = c'_n/D\right)$

$$v_0 = \frac{1}{4}\operatorname{arcsinh}\left(\frac{1}{\sqrt{10^{\frac{R}{10}}-1}}\right) \quad \kappa = \sinh(v_0) \quad \lambda = \cosh(v_0)$$

$$\alpha = \cos\left(\frac{5\pi}{8}\right) \quad \beta = \cos\left(\frac{7\pi}{8}\right) \quad \gamma = \tan\left(\pi f_c / F_s\right)$$

$$D = \gamma^4 - 2\kappa(\alpha+\beta)\gamma^3 + \left(\left(1+\sqrt{2}\right)\kappa^2 + \lambda^2\right)\gamma^2$$

$$-\kappa\left(\frac{\sqrt{2}}{2}\kappa^2(\alpha+\beta) + 2\lambda^2\left(\alpha^3+\beta^3\right)\right)\gamma + \left(\frac{\kappa^4+\lambda^4}{8} + \frac{3\kappa^2\lambda^2}{4}\right)$$

Numerator Coefficients

$b'_0 = \frac{\kappa^4+\lambda^4}{8} + \frac{3\kappa^2\lambda^2}{4}$	$b'_2 = 3\left(\frac{\kappa^4+\lambda^4}{4} + \frac{3\kappa^2\lambda^2}{2}\right)$	
$b'_1 = \frac{-\kappa^4-\lambda^4}{2} - 3\kappa^2\lambda^2$	$b'_3 = b'_1$	$b'_4 = b'_0$

Denominator Coefficients
$a'_1 = 4\gamma^4 - 4\kappa(\alpha+\beta)\gamma^3 + \kappa\left(\sqrt{2}\kappa^2(\alpha+\beta) + 4\lambda^2(\alpha^3+\beta^3)\right)\gamma$ $- \frac{\kappa^4+\lambda^4}{2} - 3\kappa^2\lambda^2$
$a'_2 = 6\gamma^4 - 2\left(\left(1+\sqrt{2}\right)\kappa^2 + \lambda^2\right)\gamma^2 + \left(3\frac{\kappa^4+\lambda^4}{4} + \frac{9\kappa^2\lambda^2}{2}\right)$
$a'_3 = 4\gamma^4 + 4\kappa(\alpha+\beta)\gamma^3 - \kappa\left(\sqrt{2}\kappa^2(\alpha+\beta) + 4\lambda^2(\alpha^3+\beta^3)\right)\gamma$ $- \frac{\kappa^4+\lambda^4}{2} - 3\kappa^2\lambda^2$
$a'_4 = \gamma^4 + 2\kappa(\alpha+\beta)\gamma^3 + \left(\left(1+\sqrt{2}\right)\kappa^2 + \lambda^2\right)\gamma^2$ $+ \kappa\left(\frac{\sqrt{2}}{2}\kappa^2(\alpha+\beta) + 2\lambda^2(\alpha^3+\beta^3)\right)\gamma + \left(\frac{\kappa^4+\lambda^4}{8} + \frac{3\kappa^2\lambda^2}{4}\right)$

The equations have again been broken up for formatting purposes.

Example: f_c = 10 kHz, F_s = 100 kHz, R = 0.25 dB. Coefficients are shown in **Table 15**; frequency response plots follow in **Figure 91**.

1st. Ord	$b_0 = -b_1 = 0.926704839684869$ $a_1 = -0.853409679369738$
2nd. Ord	$b_0 = b_2 = 0.754100705994557$ $b_1 = -1.508201411989115$ $a_1 = -1.432883331170040$ $a_2 = 0.583519492808190$
3rd. Ord	$b_0 = -b_3 = 0.555290826853105$ $b_1 = -b_2 = -1.665872480559316$ $a_1 = -1.861220016163596$ $a_2 = 1.295350071271879$ $a_3 = -0.285756527389368$
4th. Ord	$b_0 = b_4 = 0.420656954052415$ $b_2 = 2.523941724314493$ $b_1 = b_3 = -1.682627816209662$ $a_1 = -2.284592750552269$ $a_2 = 2.232156684526796$ $a_3 = -1.009693662139922$ $a_4 = 0.204068167619660$

Table 15 Coefficients: T1 Chebychev Highpass Filters

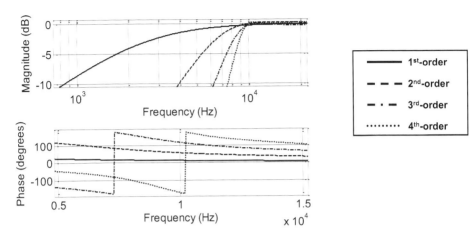

Figure 91 Frequency Responses: T1 Chebychev Highpass Filters

Taking a closer look at the passband ripple (**Figure 92**), we see that the second- and fourth-order filters hit peaks of 0.25 dB above 0 dB, and hit the cutoff frequency at 0 dB; similarly, the first- and third-order filters hit passband minima of 0.25 dB below 0 dB, and are at that point (again) at the cutoff frequency. These filters are designed to be exactly 0 dB at the Nyquist frequency (50 kHz).

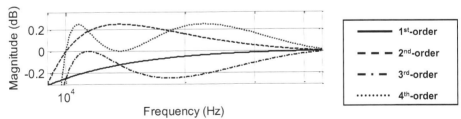

Figure 92 Passband Ripple of Type I Chebychev Highpass Filters

2.3.4.3 Type I Chebychev Bandpass Filters

Since bandpass filters generally have even orders, we show equations only for second- and forth-order filters below.

Second Order Chebychev Type I Bandpass		
Definitions:		
$v_0 = \text{arcsinh}\left(\dfrac{1}{\sqrt{10^{\frac{R}{10}}-1}}\right)$		$\kappa = \sinh(v_0)$
$\gamma = \tan\left(\dfrac{\pi f_c}{F_s}\right)$		$D = f_c\gamma^2 + \kappa BW\gamma + f_c \quad (c_n = c'_n/D)$
Numerator Coefficients			**Denominator Coefficients**
$b'_0 = \kappa BW\gamma$	$b'_1 = 0$	$b'_2 = -b'_0$	$a'_1 = 2f_c(\gamma^2 - 1)$ $a'_2 = f_c\gamma^2 - \kappa BW\gamma + f_c$

Note that the above definition for v_0 is the same as for the first-order highpass and lowpass filters. This is because the second-order bandpass filter begins with the first-order filter, as described in the **Theory** section.

This time the definition for v_0 is the same as for the second-order highpass and lowpass filters since the fourth-order bandpass filter begins with the second-order filter.

Fourth Order Chebychev Type I Bandpass
Definitions: $(c_n = c'_n/D)$
$v_0 = \dfrac{1}{2}\text{arcsinh}\left(\dfrac{1}{\sqrt{10^{\frac{R}{10}}-1}}\right) \quad \kappa = \sinh(v_0) \quad \lambda = \cosh(v_0)$
$\alpha = \cos\left(\dfrac{5\pi}{8}\right) \quad \beta = \cos\left(\dfrac{7\pi}{8}\right) \quad \gamma = \tan\left(\dfrac{\pi f_c}{F_s}\right)$
$D = 2f_c^2(\gamma^4 + 1) + 2\sqrt{2}\kappa BW f_c\gamma(\gamma^2 + 1) + (BW^2(\kappa^2 + \lambda^2) + 4f_c^2)\gamma^2$

Numerator Coefficients				
$b'_0 = BW^2\gamma^2(\kappa^2+\lambda^2)$	$b'_1 = 0$	$b'_2 = -2b'_0$	$b'_3 = 0$	$b'_4 = b'_0$
Denominator Coefficients				
$a'_1 = 8f_c^2(\gamma^4-1)+4\sqrt{2}\kappa BW f_c\gamma(\gamma^2-1)$				
$a'_2 = 12f_c^2(\gamma^4+1)-2(4f_c^2+BW^2(\lambda^2+\kappa^2))\gamma^2$				
$a'_3 = 8f_c^2(\gamma^4-1)-4\sqrt{2}\kappa BW f_c\gamma(\gamma^2-1)$				
$a'_4 = 2f_c^2(\gamma^4+1)-2\sqrt{2}\kappa BW f_c\gamma(\gamma^2+1)+(BW^2(\kappa^2+\lambda^2)+4f_c^2)\gamma^2$				

Example: f_c = 100 Hz, BW = 50 Hz (Q = 2), F_s = 1 kHz, R = 0.5 dB. Coefficients are shown in **Table 16**; frequency response plots follow in **Figure 93**.

2nd- Ord	$b_0 = -b_2 = 0.296108874776357 \quad b_1 = 0$ $a_1 = -1.138919764991264 \quad a_2 = 0.407782250447287$
4th- Ord	$b_0 = b_4 = 0.026355583328146 \quad b_2 = -0.052711166656293$ $b_1 = b_3 = 0$ $a_1 = -2.877913477824740 \quad a_2 = 3.664824339952579$ $a_3 = -2.332181396947723 \quad a_4 = 0.662719025266797$

Table 16 Coefficients: T1 Chebychev Bandpass Filters

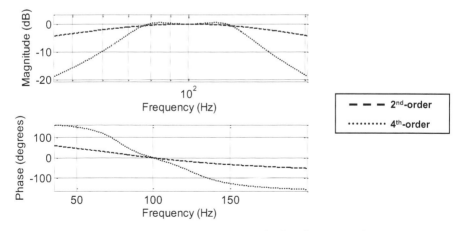

Figure 93 Frequency Responses: T1 Chebychev Bandpass Filters

Taking a closer look at the passband ripple (**Figure 94**), we see that the fourth-order filter hits a peak of 0.5 dB above 0 dB.

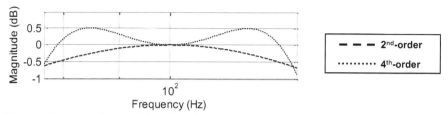

Figure 94 Passband Ripple of Type I Chebychev Bandpass Filters

2.3.4.4 Type I Chebychev Bandstop Filters

As with the bandpass case, we show only even-order equations.

Second Order Chebychev Type I Bandstop	
Definitions:	
$v_0 = \operatorname{arcsinh}\left(\dfrac{1}{\sqrt{10^{\frac{R}{10}}-1}}\right)$ $\kappa = \sinh(v_0)$ $\gamma = \tan\left(\dfrac{\pi f_c}{F_s}\right)$ $D = \kappa f_c \gamma^2 + BW\gamma + \kappa f_c$ $(c_n = c'_n / D)$	
Numerator Coefficients	**Denominator Coefficients**
$b'_0 = f_c \kappa (\gamma^2 + 1)$ $b'_2 = b'_0$	$a'_1 = 2\kappa f_c (\gamma^2 - 1)$
$b'_1 = 2 f_c \kappa (\gamma^2 - 1)$	$a'_2 = \kappa f_c \gamma^2 - BW\gamma + \kappa f_c$

Once again note that the definition for v_0 is the same as for the first-order highpass and lowpass filters. This is because the second-order bandstop filter begins with the first-order filter, as described in the **Theory** section.

Fourth Order Chebychev Type I Bandstop
Definitions: $\quad\quad\quad\quad\quad (c_n = c'_n/D)$
$v_0 = \dfrac{1}{2}\operatorname{arcsinh}\left(\dfrac{1}{\sqrt{10^{\frac{R}{10}}-1}}\right) \quad \gamma = \tan\left(\pi f_c / F_s\right)$
$\alpha = \cos\left(\dfrac{5\pi}{8}\right) = -0.38268343236509 \quad \kappa = \sinh(v_0)$
$\beta = \cos\left(\dfrac{7\pi}{8}\right) = -0.92387953251129 \quad \lambda = \cosh(v_0)$
$D = f_c^2\left(\kappa^2+\lambda^2\right)\left(\gamma^4+1\right) + 2\sqrt{2}\,BW f_c \kappa\gamma\left(\gamma^2+1\right) + 2\left(BW^2 + f_c^2\left(\kappa^2+\lambda^2\right)\right)\gamma^2$

Numerator Coefficients	
$b'_0 = f_c^2\left(\kappa^2+\lambda^2\right)\left(\gamma^4+2\gamma^2+1\right)$	$b'_3 = b'_1$
$b'_1 = 4 f_c^2\left(\kappa^2+\lambda^2\right)\left(\gamma^4-1\right)$	$b'_4 = b'_0$
$b'_2 = 2 f_c^2\left(\lambda^2+\kappa^2\right)\left(3\gamma^4-2\gamma^2+3\right)$	

Denominator Coefficients
$a'_1 = 4 f_c^2\left(\lambda^2+\kappa^2\right)\left(\gamma^4-1\right) + 4\sqrt{2}\,BW f_c \kappa\gamma\left(\gamma^2-1\right)$
$a'_2 = 6 f_c^2\left(\kappa^2+\lambda^2\right)\left(\gamma^4+1\right) - 4\left(f_c^2\left(\kappa^2+\lambda^2\right)+BW^2\right)\gamma^2$
$a'_3 = 4 f_c^2\left(\kappa^2+\lambda^2\right)\left(\gamma^4-1\right) - 4\sqrt{2}\,BW f_c \kappa\gamma\left(\gamma^2-1\right)$
$a'_4 = f_c^2\left(\kappa^2+\lambda^2\right)\left(\gamma^4+1\right) - 2\sqrt{2}\,BW f_c \kappa\gamma\left(\gamma^2+1\right) + 2\left(BW^2 + f_c^2\left(\kappa^2+\lambda^2\right)\right)\gamma^2$

Example: f_c = 1 kHz, BW = 1.5 kHz (Q = 2/3), F_s = 12 kHz, R = 0.75 dB. Coefficients are shown in **Table 17**; frequency response plots follow in **Figure 95**.

2nd-Ord	$b_0 = b_2 = 0.859983385676495$ $b_1 = -1.489534917656791$ $a_1 = -1.489534917656791$ $a_2 = 0.719966771352990$
4th-Ord	$b_0 = b_4 = 0.676058515338654$ $b_2 = 3.380292576693267$ $b_1 = b_3 = -2.341935394912262$ $a_1 = -2.771860232181701$ $a_2 = 3.228844048428209$ $a_3 = -1.912010557642823$ $a_4 = 0.503565558942366$

Table 17 Coefficients: T1 Chebychev Bandstop Filters

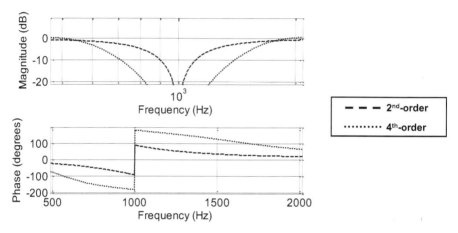

Figure 95 Frequency Responses: T1 Chebychev Bandstop Filters

Taking a closer look at the passband ripple (**Figure 96**), we see that the fourth-order filter hits a peak of 0.75 dB above 0 dB.

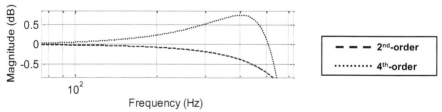

Figure 96 Passband Ripple of Type I Chebychev Bandstop Filters

2.3.5 Chebychev Type II Filters

Chebychev Type II filters differ from Type I filters in the location of the ripple. As seen in the previous section, Type I filters have ripple in the passband, while Type II filters have ripple in the stopband.

2.3.5.1 Type II Chebychev Lowpass Filters

The design equations for Chebychev Type II lowpass filters are shown below. Note the definition for v_0 differs from that of Type I.

First Order Chebychev Type II Lowpass	
Definitions: $D = \kappa + \gamma$ $\quad (c_n = c'_n/D)$	
$v_0 = \operatorname{arcsinh}\left(\sqrt{10^{\frac{R}{10}}-1}\right) \quad \kappa = \sinh(v_0) \quad \gamma = \tan\left(\pi f_c / F_s\right)$	
Numerator Coefficients	**Denominator Coefficient**
$b'_0 = \gamma \quad\quad b'_1 = b'_0$	$a'_1 = \gamma - \kappa$

Second Order Chebychev Type II Lowpass	
Definitions: $v_0 = \dfrac{1}{2}\operatorname{arcsinh}\left(\sqrt{10^{\frac{R}{10}}-1}\right) \quad \kappa = \sinh(v_0) \quad \lambda = \cosh(v_0)$	
$\gamma = \tan\left(\pi f_c / F_s\right) \quad D = 2\gamma^2 + 2\sqrt{2}\kappa\gamma + \kappa^2 + \lambda^2 \quad (c_n = c'_n/D)$	
Numerator Coefficients	**Denominator Coefficients**
$b'_0 = 2\gamma^2 + 1 \quad\quad b'_2 = b'_0$	$a'_1 = 4\gamma^2 - 2(\kappa^2 + \lambda^2)$
$b'_1 = 2(2\gamma^2 - 1)$	$a'_2 = 2\gamma^2 - 2\sqrt{2}\kappa\gamma + \kappa^2 + \lambda^2$

Third Order Chebychev Type II Lowpass	
Definitions: $\quad (c_n = c'_n/D)$	
$v_0 = \dfrac{1}{3}\operatorname{arcsinh}\left(\sqrt{10^{\frac{R}{10}}-1}\right) \quad \kappa = \sinh(v_0) \quad \lambda = \cosh(v_0) \quad \gamma = \tan\left(\pi f_c / F_s\right)$	
$D = 4\gamma^3 + 8\kappa\gamma^2 + (5\kappa^2 + 3\lambda^2)\gamma + \kappa(\kappa^2 + 3\lambda^2)$	

Numerator Coefficients			
$b'_0 = \gamma(4\gamma^2 + 3)$	$b'_1 = 3\gamma(4\gamma^2 - 1)$	$b'_2 = b'_0$	$b'_3 = b'_0$

Denominator Coefficients
$a'_1 = 12\gamma^3 + 8\kappa\gamma^2 - (5\kappa^2 + 3\lambda^2)\gamma - 3\kappa(\kappa^2 + 3\lambda^2)$
$a'_2 = 12\gamma^3 - 8\kappa\gamma^2 - (5\kappa^2 + 3\lambda^2)\gamma + 3\kappa(\kappa^2 + 3\lambda^2)$
$a'_3 = 4\gamma^3 - 8\kappa\gamma^2 + (5\kappa^2 + 3\lambda^2)\gamma - \kappa(\kappa^2 + 3\lambda^2)$

Fourth Order Chebychev Type II Lowpass
Definitions:
$v_0 = \dfrac{1}{4}\operatorname{arcsinh}\left(\sqrt{10^{\frac{R}{10}}-1}\right)$ $\quad\quad \alpha = \cos\left(\dfrac{5\pi}{8}\right) = -0.38268343236509$ $\kappa = \sinh(v_0) \quad \lambda = \cosh(v_0) \quad\quad \beta = \cos\left(\dfrac{7\pi}{8}\right) = -0.92387953251129$ $\gamma = \tan\left(\dfrac{\pi f_c}{F_s}\right) \quad\quad (c_n = c'_n/D)$ $D = 8\gamma^4 - 16\kappa(\alpha+\beta)\gamma^3 + 8\left(\left(1+\sqrt{2}\right)\kappa^2 + \lambda^2\right)\gamma^2$ $\quad -4\kappa\left(\sqrt{2}\kappa^2(\alpha+\beta) + 4\lambda^2(\alpha^3+\beta^3)\right)\gamma + \kappa^4 + \lambda^4 + 6\kappa^2\lambda^2$
Numerator Coefficients

$b'_0 = 8\gamma^4 + 8\gamma^2 + 1$		$b'_1 = 4(8\gamma^4 - 1)$
$b'_2 = 2(24\gamma^4 - 8\gamma^2 + 3)$	$b_3 = b'_1$	$b_4 = b'_0$

Denominator Coefficients
$a'_1 = 32\gamma^4 - 32\kappa(\alpha+\beta)\gamma^3 + 8\kappa\left(\sqrt{2}\kappa^2(\alpha+\beta) + 4\lambda^2(\alpha^3+\beta^3)\right)\gamma$ $\quad\quad - 4\left(\kappa^4 + \lambda^4 + 6\kappa^2\lambda^2\right)$
$a'_2 = 48\gamma^4 - 16\left(\left(1+\sqrt{2}\right)\kappa^2 + \lambda^2\right)\gamma^2 + 6\left(\kappa^4 + \lambda^4 + 6\kappa^2\lambda^2\right)$
$a'_3 = 32\gamma^4 + 32\kappa(\alpha+\beta)\gamma^3 - 8\kappa\left(\sqrt{2}\kappa^2(\alpha+\beta) + 4\lambda^2(\alpha^3+\beta^3)\right)\gamma$ $\quad\quad - 4\left(\kappa^4 + \lambda^4 + 6\kappa^2\lambda^2\right)$
$a'_4 = 8\gamma^4 + 16\kappa(\alpha+\beta)\gamma^3 + 8\left(\left(1+\sqrt{2}\right)\kappa^2 + \lambda^2\right)\gamma^2$ $\quad +4\kappa\left(\sqrt{2}\kappa^2(\alpha+\beta) + 4\lambda^2(\alpha^3+\beta^3)\right)\gamma + \kappa^4 + \lambda^4 + 6\kappa^2\lambda^2$

Example: f_c = 500 Hz, F_s = 40 kHz, R = 25 dB. Note that ripple in this case refers to how near 0 dB the stopband can rise. Coefficients are shown in **Table 18**; frequency response plots follow in **Figure 97**.

1st-Ord	$b_0 = b_1 = 0.002208060510377$
	$a_1 = -0.995583878979247$
2nd-Ord	$b_0 = b_2 = 0.055395324006081 \quad b_1 = -0.110108638030122$
	$a_1 = -1.963762255315444 \quad a_2 = 0.964444265297486$
3rd-Ord	$b_0 = b_3 = 0.006353339726661 \quad b_1 = b_2 = -0.006301139276602$
	$a_1 = -2.908025164218774 \quad a_2 = 2.820170861929837$
	$a_3 = -0.912041296810946$
4th-Ord	$b_0 = b_4 = 0.052615521821962 \quad b_2 = 0.310558354464462$
	$b_1 = b_3 = -0.207886772435531$
	$a_1 = -3.842357386493769 \quad a_2 = 5.539515238521860$
	$a_3 = -3.551390985146191 \quad a_4 = 0.854248986355424$

Table 18 Coefficients: T2 Chebychev Lowpass Filters

Notice that the filters for all four orders are down to the ripple value at the cutoff frequency, and that the higher order filters bounce back up to that value after hitting nulls. Due to this characteristic, Type II filters may be less popular than some others. However, there are cases where this characteristic is a good fit to the problem at hand.

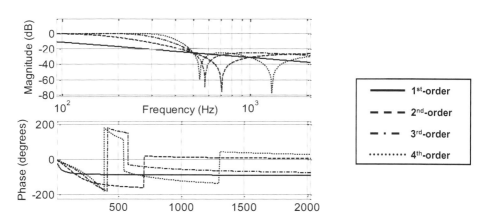

Figure 97 Frequency Responses: T2 Chebychev Lowpass Filters

109

2.3.5.2 Type II Chebychev Highpass Filters

Design equations for Type II Chebychev highpass filters are provided below.

First Order Chebychev Type II Highpass	
Definitions:	
$v_0 = \operatorname{arcsinh}\left(\sqrt{10^{\frac{R}{10}}-1}\right)$ $\quad \kappa = \sinh(v_0) \quad \gamma = \tan\left(\pi f_c / F_s\right)$	
$D = \kappa\gamma + 1 \qquad (c_n = c'_n/D)$	
Numerator Coefficients	**Denominator Coefficient**
$b'_0 = 1 \qquad b'_1 = -1$	$a'_1 = \kappa\gamma - 1$

Second Order Chebychev Type II Highpass	
Definitions: $v_0 = \dfrac{1}{2}\operatorname{arcsinh}\left(\sqrt{10^{\frac{R}{10}}-1}\right) \quad \kappa = \sinh(v_0) \quad \lambda = \cosh(v_0)$	
$\gamma = \tan\left(\pi f_c / F_s\right) \quad D = (\kappa^2 + \lambda^2)\gamma^2 + 2\sqrt{2}\kappa\gamma + 2 \quad (c_n = c'_n/D)$	
Numerator Coefficients	**Denominator Coefficients**
$b'_0 = \gamma^2 + 2$	$a'_1 = 2\left((\kappa^2 + \lambda^2)\gamma^2 - 2\right)$
$b'_1 = 2(\gamma^2 - 2) \qquad b'_2 = b'_0$	$a'_2 = (\kappa^2 + \lambda^2)\gamma^2 - 2\sqrt{2}\kappa\gamma + 2$

Third Order Chebychev Type II Highpass			
Definitions: $\qquad\qquad (c_n = c'_n/D)$			
$v_0 = \dfrac{1}{3}\operatorname{arcsinh}\left(\sqrt{10^{\frac{R}{10}}-1}\right) \quad \kappa = \sinh(v_0) \quad \lambda = \cosh(v_0)$			
$\gamma = \tan\left(\pi f_c / F_s\right) \quad D = \kappa(\kappa^2 + 3\lambda^2)\gamma^3 + (5\kappa^2 + 3\lambda^2)\gamma^2 + 8\kappa\gamma + 4$			
Numerator Coefficients			
$b'_0 = 3\gamma^2 + 4$	$b'_1 = 3(\gamma^2 - 4)$	$b'_2 = -b'_1$	$b'_3 = -b'_0$

Denominator Coefficients
$a'_1 = 3\kappa(\kappa^2 + 3\lambda^2)\gamma^3 + (5\kappa^2 + 3\lambda^2)\gamma^2 - 8\kappa\gamma - 12$
$a'_2 = 3\kappa(\kappa^2 + 3\lambda^2)\gamma^3 - (5\kappa^2 + 3\lambda^2)\gamma^2 - 8\kappa\gamma + 12$
$a'_3 = \kappa(\kappa^2 + 3\lambda^2)\gamma^3 - (5\kappa^2 + 3\lambda^2)\gamma^2 + 8\kappa\gamma - 4$

Fourth Order Chebychev Type II Highpass
Definitions: $\quad (c_n = c'_n/D)$
$v_0 = \dfrac{1}{4}\operatorname{arcsinh}\left(\sqrt{10^{\frac{R}{10}} - 1}\right) \quad \kappa = \sinh(v_0) \quad \lambda = \cosh(v_0)$
$\alpha = \cos\left(\dfrac{5\pi}{8}\right) \quad \beta = \cos\left(\dfrac{7\pi}{8}\right) \quad \gamma = \tan\left(\pi f_c / F_s\right) \quad (c_n = c'_n/D)$
$D = (\kappa^4 + \lambda^4 + 6\kappa^2\lambda^2)\gamma^4 - 4\kappa(\sqrt{2}\kappa^2(\alpha+\beta) + 4\lambda^2(\alpha^3+\beta^3))\gamma^3$ $\qquad + 8((1+\sqrt{2})\kappa^2 + \lambda^2)\gamma^2 - 16\kappa(\alpha+\beta)\gamma + 8$

Numerator Coefficients		
$b'_0 = \gamma^4 + 8\gamma^2 + 8$	$b'_1 = 4(\gamma^4 - 8)$	
$b'_2 = 2(3\gamma^4 - 8\gamma^2 + 24)$	$b'_3 = b'_1$	$b'_4 = b'_0$

Denominator Coefficients
$a'_1 = 4(\kappa^4 + \lambda^4 + 6\kappa^2\lambda^2)\gamma^4 - 8\kappa(\sqrt{2}\kappa^2(\alpha+\beta) + 4\lambda^2(\alpha^3+\beta^3))\gamma^3$ $\qquad + 32(\kappa(\alpha+\beta)\gamma - 1)$
$a'_2 = 6(\kappa^4 + \lambda^4 + 6\kappa^2\lambda^2)\gamma^4 - 16((1+\sqrt{2})\kappa^2 + \lambda^2)\gamma^2 + 48$
$a'_3 = 4(\kappa^4 + \lambda^4 + 6\kappa^2\lambda^2)\gamma^4 + 8\kappa(\sqrt{2}\kappa^2(\alpha+\beta) + 4\lambda^2(\alpha^3+\beta^3))\gamma^3$ $\qquad - 32(\kappa(\alpha+\beta)\gamma + 1)$
$a'_4 = (\kappa^4 + \lambda^4 + 6\kappa^2\lambda^2)\gamma^4 + 4\kappa(\sqrt{2}\kappa^2(\alpha+\beta) + 4\lambda^2(\alpha^3+\beta^3))\gamma^3$ $\qquad + 8((1+\sqrt{2})\kappa^2 + \lambda^2)\gamma^2 + 16\kappa(\alpha+\beta)\gamma + 8$

Example: f_c = 800 Hz, F_s = 10 kHz, R = 15 dB. Coefficients are shown in **Table 19**; frequency response plots follow in **Figure 98**.

1st-Ord	$b_0 = -b_1 = 0.413080695979987$ $a_1 = 0.173838608040026$
2nd-Ord	$b_0 = b_2 = 0.594531269285680 \quad b_1 = -1.113176338034875$ $a_1 = -0.937749705436379 \quad a_2 = 0.364489171169856$
3rd-Ord	$b_0 = -b_3 = 0.641099081458040 \quad b_1 = -b_2 = -1.802479693162262$ $a_1 = -1.975554797914375 \quad a_2 = 1.500799356265122$ $a_3 = -0.410803395061107$
4th-Ord	$b_0 = b_4 = 0.658023629920747 \quad b_2 = 3.622735112177787$ $b_1 = b_3 = -2.466709679339644$ $a_1 = -2.953683133635730 \quad a_2 = 3.518181547452052$ $a_3 = -1.967338511570427 \quad a_4 = 0.432998538040358$

Table 19 Coefficients: T2 Chebychev Highpass Filters

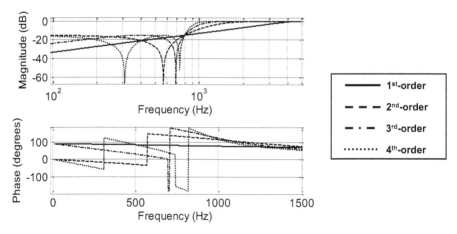

Figure 98 Frequency Responses: T2 Chebychev Highpass Filters

2.3.5.3 Type II Chebychev Bandpass Filters

As with the other bandpass filters we have considered, the Type II Chebychev variety are shown below for even orders only.

Second Order Chebychev Type II Bandpass	
Definitions:	
$v_0 = \operatorname{arcsinh}\left(\sqrt{10^{\frac{R}{10}}-1}\right)$ $\kappa = \sinh(v_0)$ $\gamma = \tan\left(\pi f_c / F_s\right)$	$D = \kappa f_c\left(\gamma^2+1\right)+BW\gamma$ $(c_n = c'_n/D)$
Numerator Coefficients	Denominator Coefficients
$b'_0 = BW\gamma$	$a'_1 = 2\kappa f_c\left(\gamma^2-1\right)$
$b'_1 = 0 \quad b'_2 = -b'_0$	$a'_2 = \kappa f_c\left(\gamma^2+1\right)-BW\gamma$

Once again, the definition for v_0 is the same as for the first-order highpass and lowpass filters.

Fourth Order Chebychev Type II Bandpass	
Definitions: $v_0 = \frac{1}{2}\operatorname{arcsinh}\left(\sqrt{10^{\frac{R}{10}}-1}\right) \quad \kappa = \sinh(v_0) \quad \lambda = \cosh(v_0)$ $\alpha = \cos\left(\frac{5\pi}{8}\right) \quad \beta = \cos\left(\frac{7\pi}{8}\right) \quad \gamma = \tan\left(\pi f_c/F_s\right)$ $D = f_c^2\left(\kappa^2+\lambda^2\right)\gamma^4 + 2\sqrt{2}\kappa f_c BW\left(\gamma^2+1\right)\gamma \quad (c_n = c'_n/D)$ $\quad + 2\left(f_c^2\left(\kappa^2+\lambda^2\right)+BW^2\right)\gamma^2 + f_c^2\left(\kappa^2+\lambda^2\right)$	
Numerator Coefficients	
$b'_0 = f_c^2\gamma^4 + 2\left(f_c^2+BW^2\right)\gamma^2 + f_c^2$	$b'_1 = 4f_c^2\left(\gamma^4-1\right)$
$b'_2 = 2\left(3f_c^2\gamma^4 - 2\left(f_c^2+BW^2\right)\gamma^2 + 3f_c^2\right)$	$b'_3 = b'_1 \quad b'_4 = b'_0$

Digital Filters for Everyone

Denominator Coefficients
$a'_1 = 4f_c\left(f_c(\kappa^2 + \lambda^2)\gamma^4 + \sqrt{2}\kappa BW(\gamma^2 - 1)\gamma - f_c(\kappa^2 + \lambda^2)\right)$
$a'_2 = 2\left(3f_c^2(\kappa^2 + \lambda^2)\gamma^4 - 2\left(f_c^2(\kappa^2 + \lambda^2) + BW^2\right)\gamma^2 + 3f_c^2(\kappa^2 + \lambda^2)\right)$
$a'_3 = 4f_c\left(f_c(\kappa^2 + \lambda^2)\gamma^4 - \sqrt{2}\kappa BW(\gamma^2 - 1)\gamma - f_c(\kappa^2 + \lambda^2)\right)$
$a'_4 = f_c^2(\kappa^2 + \lambda^2)\gamma^4 - 2\sqrt{2}\kappa f_c BW(\gamma^2 + 1)\gamma$ $+ 2\left(f_c^2(\kappa^2 + \lambda^2) + BW^2\right)\gamma^2 + f_c^2(\kappa^2 + \lambda^2)$

Example: f_c = 600 Hz, BW = 250 Hz (Q = 12/5), F_s = 10 kHz, R = 20 dB. Coefficients are shown in **Table 20**; frequency response plots follow in **Figure 99**.

2nd-Ord	$b_0 = -b_2 = 0.007648940584362 \quad b_1 = 0$ $a_1 = -1.845329361581910 \quad a_2 = 0.984702118831275$
4th-Ord	$b_0 = b_4 = 0.096616818235540 \quad b_2 = 0.518950740493448$ $b_1 = b_3 = -0.355150366484304$ $a_1 = -3.633215895966685 \quad a_2 = 5.209727505536478$ $a_3 = -3.469791433719392 \quad a_4 = 0.912116264108805$

Table 20 Coefficients: T2 Chebychev Bandpass

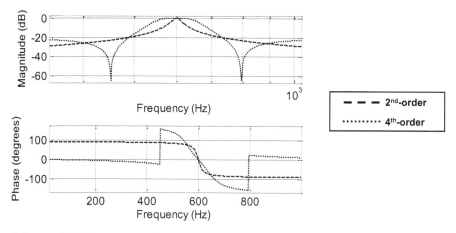

Figure 99 Frequency Responses: T2 Chebychev Bandpass Filters

2.3.5.4 Type II Chebychev Bandstop Filters

Design equations for even-order Type II Chebychev bandstop filters are shown below.

Second Order Chebychev Type II Bandstop	
Definitions:	
$v_0 = \operatorname{arcsinh}\left(\sqrt{10^{\frac{R}{10}}-1}\right) \quad \kappa = \sinh(v_0) \quad \gamma = \tan\left(\pi f_c / F_s\right)$	
$D = f_c \gamma^2 + BW \kappa \gamma + f_c \qquad (c_n = c'_n / D)$	
Numerator Coefficients	Denominator Coefficients
$b'_0 = f_c(\gamma^2 + 1)$	$a'_1 = b'_1$
$b'_1 = 2f_c(\gamma^2 - 1) \qquad b'_2 = b'_0$	$a'_2 = f_c \gamma^2 - BW \kappa \gamma + f_c$

Fourth Order Chebychev Type II Bandstop		
Definitions:		
$v_0 = \dfrac{1}{2}\operatorname{arcsinh}\left(\sqrt{10^{\frac{R}{10}}-1}\right) \quad \kappa = \sinh(v_0) \quad \lambda = \cosh(v_0)$		
$\alpha = \cos\left(\dfrac{5\pi}{8}\right) \quad \beta = \cos\left(\dfrac{7\pi}{8}\right) \quad \gamma = \tan\left(\pi f_c / F_s\right) \quad (c_n = c'_n / D)$		
$D = 2f_c^2 \gamma^4 + 2\sqrt{2}\kappa f_c BW(\gamma^2+1)\gamma + \left(4f_c^2 + BW^2(\kappa^2+\lambda^2)\right)\gamma^2 + 2f_c^2$		
Numerator Coefficients		
$b'_0 = 2f_c^2\gamma^4 + \left(4f_c^2 + BW^2\right)\gamma^2 + 2f_c^2$	$b'_1 = 8f_c^2(\gamma^4 - 1)$	
$b'_2 = 2\left(6f_c^2\gamma^4 - \left(4f_c^2 + BW^2\right)\gamma^2 + 6f_c^2\right)$	$b'_3 = b'_1$	$b'_4 = b'_0$
Denominator Coefficients		
$a'_1 = 4f_c\left(2f_c\gamma^4 + \sqrt{2}\kappa BW(\gamma^2-1)\gamma - 2f_c\right)$		
$a'_2 = 2\left(6f_c^2\gamma^4 - \left(4f_c^2 + BW^2(\kappa^2+\lambda^2)\right)\gamma^2 + 6f_c^2\right)$		
$a'_3 = 4f_c\left(2f_c\gamma^4 - \sqrt{2}\kappa BW(\gamma^2-1)\gamma - 2f_c\right)$		
$a'_4 = 2f_c^2\gamma^4 - 2\sqrt{2}\kappa f_c BW(\gamma^2+1)\gamma + \left(4f_c^2 + BW^2(\kappa^2+\lambda^2)\right)\gamma^2 + 2f_c^2$		

Once again the definition for v_0 is the same as for the second-order highpass and lowpass filters because the fourth-order bandstop filter begins with the second-order filter.

Example: f_c = 1.3 kHz, BW = 2 kHz, F_s = 100 kHz, R = 40 dB. Coefficients are shown in **Table 21**; frequency response plots follow in **Figure 100**.

2nd- Ord	$b_0 = b_2$ = 0.137440285989691 b_1 = -0.273964100370357 a_1 = -0.273964100370357 a_2 = -0.725119428020617
4th- Ord	$b_0 = b_4$ = 0.550101216954798 b_2 = 3.277337427692900 $b_1 = b_3$ = -2.188757724951371 a_1 = -2.872168984331746 a_2 = 3.063237285541164 a_3 = -1.505346465570997 a_4 = 0.314302576061333

Table 21 Coefficients: T2 Chebychev Bandstop Filters

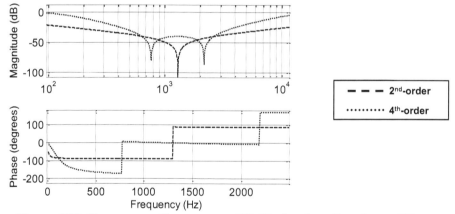

Figure 100 Frequency Responses: T2 Chebychev Bandstop Filters

2.3.6 Variable Q Filters

Aside from Type I Chebychev filters, the filters we have considered so far are smooth in the passband. In some cases, it is useful to set the passband peaking of the filter. I have done this, for instance, in audio equalization situations where I needed a highpass filter to keep energy below the reasonable response of the speaker from causing distortion or heat in the speaker coil, but where we still wanted to get all the bass response the speaker could muster. A filter type that allows this adjustable peaking is the variable Q filter. The parameter Q is somewhat similar to the Q used in bandpass and bandstop filters in that it determines the width of the peaking portion of the filter. However, in the case of a variable Q filter, Q also determines the linear gain value at the cutoff frequency.

2.3.6.1 Variable Q Lowpass Filters

Equations for variable Q, second-order, lowpass filters are provided.

Second Order Variable Q Lowpass			
Definitions: $\gamma = \tan\left(\pi f_c / F_s\right)$ $\quad D = Q\gamma^2 + \gamma + Q \quad (c_n = c'_n/D)$			
Numerator Coefficients		**Denominator Coefficients**	
$b'_0 = Q\gamma^2$ $\quad b'_1 = 2b'_0 \quad b'_2 = b'_0$		$a'_1 = 2Q(\gamma^2 - 1)$	$a'_2 = Q\gamma^2 - \gamma + Q$

Butterworth filters have Qs of $\frac{\sqrt{2}}{2}$ and Linkwitz-Reilly filters have Qs of $\frac{1}{2}$. Therefore, as also evident in the plots below, if these values are substituted for Q, the equations for those filters will be the result.

Example: f_c = 100 Hz, F_s = 10 kHz. Let us consider the following values of Q: 0.25, 0.5 (Linkwitz-Reilly), $\frac{\sqrt{2}}{2}$ (Butterworth), 1, 2, and 4. Coefficients for the 0.25 and 4 cases are shown in **Table 22**; frequency response plots follow in **Figure 101**.

Q=0.25	$b_0 = b_2$ = 0.000876556864079 b_1 = 0.001753113728158 a_1 = -1.773353839121109 a_2 = 0.776860066577426
Q=4	$b_0 = b_2$ = 0.000978952171434 b_1 = 0.001957904342869 a_1 = -1.980508809719708 a_2 = 0.984424618405445

Table 22 Coefficients for Variable Q Lowpass Filters

Note that higher values of Q correspond to the higher peaks.

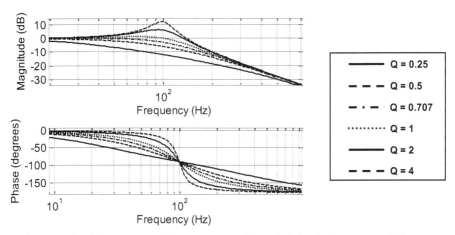

Figure 101 Frequency Responses of Variable Q Lowpass Filters

2.3.6.2 Variable Q Highpass Filters

Equations for variable Q, second-order, highpass filters are provided.

As examples, let us consider variable Q highpass filters with cutoff frequencies of 1 kHz, each operating at a sample rate of 20 kHz. Let us consider the following values of Q: 0.25, 0.5 (Linkwitz-Reilly), $\frac{\sqrt{2}}{2}$ (Butterworth), 1, 2, and 4. The coefficients for the 0.25 and 5 cases are shown in **Table 23** below and the frequency response plots follow in **Figure 102**.

Second Order Variable Q Highpass			
Definitions: $\gamma = \tan\left(\pi f_c / F_s\right)$ $D = Q\gamma^2 + \gamma + Q$ $(c_n = c'_n/D)$			
Numerator Coefficients		Denominator Coefficients	
$b'_0 = Q$ $b'_1 = -2b'_0$ $b'_2 = b'_0$		$a'_1 = 2Q(\gamma^2 - 1)$	$a'_2 = Q\gamma^2 - \gamma + Q$

Q=0.25	$b_0 = b_2 = 0.602909620521184$ $b_1 = -1.205819241042368$ $a_1 = -1.175570504584946$ $a_2 = 0.236067977499790$
Q=4	$b_0 = b_2 = 0.939247816012885$ $b_1 = -1.878495632025770$ $a_1 = -1.831372383884161$ $a_2 = 0.925618880167378$

Table 23 Coefficients for Variable Q Highpass Filters

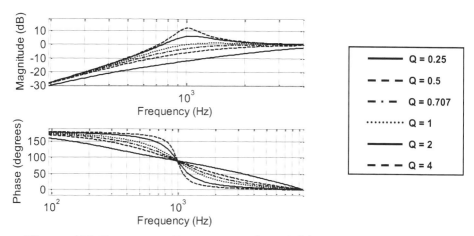

Figure 102 Frequency Responses of Variable Q Highpass Filters

Once again, the higher values of Q correspond to the higher peaks.

2.3.6.3 Variable Q Bandpass Filters

Since the variable Q lowpass filter is second order, the resulting bandpass filters will be fourth order. The equations for variable Q, fourth-order, bandpass filters are shown below.

Example: f_c = 2.5 kHz, BW = 2.5 kHz, F_s = 120 kHz, Q: 0.25, 0.5 (Linkwitz-Reilly), $\frac{\sqrt{2}}{2}$ (Butterworth), 1, 2, and 4. Coefficients for the 0.25 and 4 cases are shown in **Table 24**; frequency response plots follow in **Figure 103**.

Fourth Order Variable Q Bandpass				
Definitions: $\gamma = \tan\left(\pi f_c / F_s\right)$ $\quad (c_n = c'_n/D)$				
$D = Qf_c^2\gamma^4 + f_c BW(\gamma^2+1)\gamma + Q(2f_c^2 + BW^2)\gamma^2 + Qf_c^2$				
Numerator Coefficients				
$b'_0 = QBW^2\gamma^2$	$b'_1 = 0$	$b'_2 = -2b'_0$	$b'_3 = 0$	$b'_4 = b'_0$

Denominator Coefficients
$a_1' = 2f_c\left(2Qf_c\gamma^4 + BW(\gamma^2-1)\gamma - 2Qf_c\right)$
$a_2' = 2Q\left(3f_c^2\gamma^4 - (2f_c^2 + BW^2)\gamma^2 + 3f_c^2\right)$
$a_3' = 2f_c\left(2Qf_c\gamma^4 - BW(\gamma^2-1)\gamma - 2Qf_c\right)$
$a_4' = Qf_c^2\gamma^4 - f_cBW(\gamma^2+1)\gamma + Q(2f_c^2 + BW^2)\gamma^2 + Qf_c^2$

$Q=0.25$	$b_0 = b_4 = 0.003366183890863$ $b_1 = b_3 = 0 \quad b_2 = -0.006732367781726$ $a_1 = -3.543330620479733 \quad a_2 = 4.681323435488547$ $a_3 = -2.725132056216996 \quad a_4 = 0.587370616289752$
$Q=4$	$b_0 = b_4 = 0.004173403740998 \quad b_1 = b_3 = 0$ $b_2 = -0.008346807481996$ $a_1 = -3.917528498091689 \quad a_2 = 5.803917246328060$ $a_3 = -3.854128200832189 \quad a_4 = 0.968026312037339$

Table 24 Coefficients for Variable Q Bandpass Filters

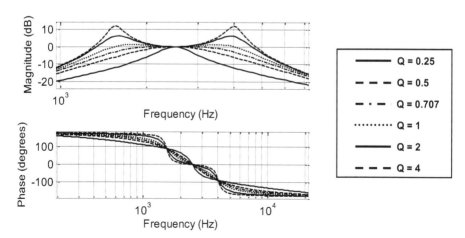

Figure 103 Frequency Responses of Variable Q Bandpass Filters

2.3.6.4 Variable Q Bandstop Filters

Equations for variable Q, fourth-order, bandstop filters are shown.

Fourth Order Variable Q Bandstop
Definitions: $\gamma = \tan\left(\pi f_c / F_s\right)$ $\qquad (c_n = c'_n/D)$
$D = Qf_c^2\gamma^4 - BW f_c(\gamma^2+1)\gamma + Q(2f_c^2+BW^2)\gamma^2 + Qf_c^2$
Numerator Coefficients
$b'_0 = Qf_c^2(\gamma^4+2\gamma^2+1)\qquad\qquad b'_1 = 4Qf_c^2(\gamma^4-1)$
$b'_2 = 2Qf_c^2(3\gamma^4-2\gamma^2+3)\qquad b'_3 = b'_1 \qquad b'_4 = b'_0$
Denominator Coefficients
$a'_1 = 2f_c\left(2Qf_c\gamma^4 - BW(\gamma^2-1)\gamma - 2Qf_c\right)$
$a'_2 = 2Q\left(3f_c^2\gamma^4 - (2f_c^2+BW^2)\gamma^2 + 3f_c^2\right)$
$a'_3 = 2f_c\left(2Qf_c\gamma^4 + BW(\gamma^2-1)\gamma - 2Qf_c\right)$
$a'_4 = Qf_c^2\gamma^4 + BW f_c(\gamma^2+1)\gamma + Q(2f_c^2+BW^2)\gamma^2 + Qf_c^2$

Example: f_c = 100 Hz, BW = 1 kHz, F_s = 10 kHz, Q: 0.25, 0.5 (Linkwitz-Reilly), $\frac{\sqrt{2}}{2}$ (Butterworth), 1, 2, and 4. Coefficients for the 0.25 and 4 cases are shown in **Table 25**; frequency response plots follow in **Figure 104**.

Q=0.25	$b_0 = b_4 =$ -6.359536780055731 $\quad b_2 =$ -38.056926986309783
	$b_1 = b_3 =$ 25.387950747673145
	$a_1 =$ 9.446724575362977 $\quad a_2 =$ -36.803255811002202
	$a_3 =$ 41.329176919983318 $\quad a_4 =$ -14.972744735419047
Q=4	$b_0 = b_4 =$ 0.980317110553384 $\quad b_2 =$ 5.866442476873822
	$b_1 = b_3 =$ -3.913530714671401
	$a_1 =$ -3.759947822941312 $\quad a_2 =$ 5.673190146292785
	$a_3 =$ -4.067113606401489 $\quad a_4 =$ 1.153886551687806

Table 25 Coefficients for Variable Q Bandstop Filters

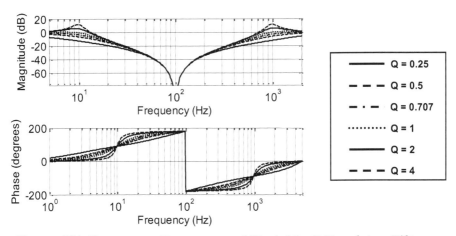

Figure 104 Frequency Responses of Variable Q Bandstop Filters

2.3.7 Allpass Filters [References 8, 26, 27, 34]

As their name implies, allpass filters pass all frequencies; their magnitude responses are identically one across the entire spectrum. However, they are used for the effect they can have on phase. In the **Digital Filter Applications** chapter of **Part 1**, we showed how to use these filters for phase equalization.

The equations for first- and second-order allpass filters are shown.

First Order Allpass	
Definitions: $\gamma = \tan\left(\pi f_c / F_s\right)$ $\quad D = \gamma + 1 \quad (c_n = c'_n / D)$	
Numerator Coefficients	Denominator Coefficient
$b'_0 = \gamma - 1 \quad\quad b'_1 = D \;\; (b_1 = 1)$	$a'_1 = b'_0$

Second Order Allpass			
Definitions:			
$\alpha = \tan\left(\pi BW / F_s\right) \quad \beta = -\cos\left(2\pi f_c / F_s\right) \quad D = 1 + \alpha \quad (c_n = c'_n / D)$			
Numerator Coefficients			Denominator Coefficients
$b'_0 = 1 - \alpha$	$b'_1 = 2\beta$	$b'_2 = D \;\; (b_2 = 1)$	$a'_1 = b'_1 \quad\quad a'_2 = b'_0$

Note that definitions of α and β are different than have been used for other filters.

Example: f_c = 1 kHz, F_s = 50 kHz. For the case of the second order allpass, let us use the same critical frequency and plot three different bandwidths: 666.67 Hz, 800 Hz, and 1 kHz. Coefficients are shown in **Table 26** (1 kHz bandwidth for second-order case); frequency response plots follow in **Figure 105**.

1st-Ord	$b_0 = -0.881618592363189 \quad b_1 = 1$
	$a_1 = -0.881618592363189$
2nd-Ord	$b_0 = 0.881618592363189 \quad b_1 = -1.866781467750174 \quad b_2 = 1$
	$a_1 = -1.866781467750174 \quad a_2 = 0.881618592363189$

Table 26 Coefficients for Allpass Filters

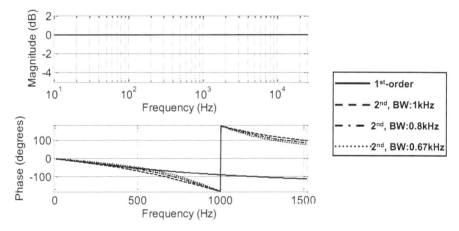

Figure 105 Frequency Responses of Allpass Filters

As promised, the magnitude responses are flat for these filters, though their phase responses are nontrivial. The associated group delays are shown in **Figure 106**. Notice that the first-order allpass filter has a shelf-like response, while the second-order filters have bell-shaped responses which are taller for narrower bandwidths. These filters can be strategically placed to equalize group delay in systems or in conjunction with other filters.

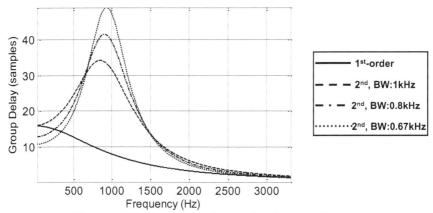

Figure 106 Group Delays of Allpass Filters

2.3.8 Equalization Filters [References 8, 23, 26, 27]

Commonly found in audio systems, equalization filters are used to correct non-flat magnitude responses. They have a bell-shaped characteristic which is typically something other than 0 dB at the center, and 0 dB on either end.

The equations for second-order equalization filters are shown below. Note that the linear gain, g, is specified. Recall that the gain in dB is $20 \cdot \log_{10}(g)$.

g	dB*	g	dB*
1	0	$\sqrt{2}$	-3
$\sqrt{2}$	3	2	
2	6	0.5	-6
4	12	0.25	-12

*approximate

Note that for these filters the design equations are different depending upon the value of g. For gains less than 1, the left equations are used; for gains of 1 or greater, use the right.

Second Order Equalization			
Definitions: $\alpha = \tan\left(\pi BW / F_s\right)$ $\beta = -\cos\left(2\pi f_c / F_s\right)$			
Numerator Coefficients		Denominator Coefficients	
$g < 1$	$g \geq 1$	$g < 1$	$g \geq 1$
$D = \alpha + g$	$D = \alpha + 1$	$D = \alpha + g$	$D = \alpha + 1$
$b'_0 = g + \alpha g$	$b'_0 = 1 + \alpha g$	$a'_1 = 2\beta g$	$a'_1 = 2\beta$
$b'_1 = 2\beta g$	$b'_1 = 2\beta$	$a'_2 = g - \alpha$	$a'_2 = 1 - \alpha$
$b'_2 = g - \alpha g$	$b'_2 = 1 - \alpha g$	$(c_n = c'_n / D)$	

Examples: F_s = 44.1 kHz: **1)** f_c = 100 Hz, BW = 100 Hz, g = 2; **2)** f_c = 100 Hz, BW = 100 Hz, g = 0.5; **3)** f_c = 400 Hz, BW = 100 Hz, g = 4; **4)** f_c = 400 Hz, BW = 100 Hz, g = 0.25; **5)** f_c = 2.5 kHz, BW = 5 kHz, g = $\sqrt{2}$; **6)** f_c = 2.5 kHz, BW = 5 kHz, g = $\frac{\sqrt{2}}{2}$. Frequency response plots are shown in **Figure 107**. Coefficients for the first two filters follow in **Table 27**.

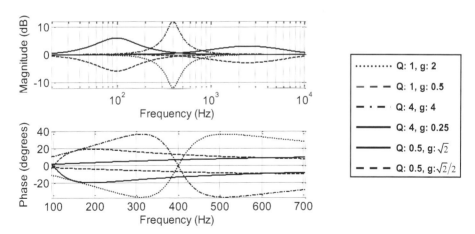

Figure 107 Frequency Responses of Equalization Filters

f_c=100 BW=100 g=2	$b_0 = 1.007073522215302$ $b_1 = -1.985651401160246$ $b_2 = 0.978779433354096$ $a_1 = -1.985651401160246$ $a_2 = 0.985852955569397$
f_c=100 BW=100 g=0.5	$b_0 = 0.992976161065439$ $b_1 = -1.971704505538310$ $b_2 = 0.978928483196317$ $a_1 = -1.971704505538310$ $a_2 = 0.971904644261756$

Table 27 Coefficients for Equalization Filters

2.3.9 Notch Filters [Reference 27]

Notch filters are essentially just bandstop filters although typically a notch filter is very narrow, while a bandstop filter can be narrow or wide. Along with the various bandstop filters in previous sections, in this section we present a simple second-order notch.

Second Order Notch	
Definitions:	
$\alpha = \tan\left(\pi BW/F_s\right)$ $\beta = -\cos\left(2\pi f_c/F_s\right)$	$D = \alpha + 1$ $(c_n = c'_n/D)$
Numerator Coefficients	Denominator Coefficients
$b'_0 = 1$ $b'_1 = 2\beta$ $b'_2 = 1$	$a'_1 = b'_1$ $a'_2 = 1 - \alpha$

Examples: F_s = 12 kHz: **1)** fc = 100 Hz, BW = 100 Hz; **2)** fc = 100 Hz, BW = 50 Hz; **3)** fc = 100 Hz, BW = 200 Hz; **4)** fc = 400 Hz, BW = 100 Hz. Coefficients for the first two cases are shown in **Table 28**; frequency response plots follow in **Figure 108**.

f_c=100 BW=100	$b_0 = b_2 = 0.974482283357440$ $b_1 = -1.946293578511630$ $a_1 = -1.946293578511630$ $a_2 = 0.948964566714880$
f_c=100 BW=50	$b_0 = b_2 = 0.987078435460841$ $b_1 = -1.971451357541064$ $a_1 = -1.971451357541064$ $a_2 = 0.974156870921681$

Table 28 Coefficients for Notch Filters

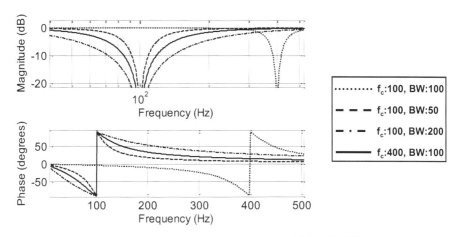

Figure 108 Frequency Responses of Notch Filters

2.3.10 Shelf Filters [References 20, 21, 26]

Shelf filters, which have a non-unity gain in a region, emphasize or deemphasize signals at the higher or lower frequencies. They differ from high- and lowpass filters in that they have finite gains in both regions, rather than going to 0 in a stopband. In this section we show how to design first- and second-order shelf filters.

Shelf filters are often associated with audio systems, so one that is active at lower-frequencies is often referred to as a bass shelf, while one active at higher-frequencies is called a treble shelf. Of course, these filters can also be used in non-audio applications.

First Order Shelf				Definition: $\gamma = \tan\left(\pi f_c / F_s\right)$		
Low			$c_n = c'_n / D$	High		
$g > 1$	$D = \gamma + 1$	$g\gamma + 1$	b'_0	$\gamma + g$	$D = \gamma + 1$	$g > 1$
		$g\gamma - 1$	b'_1	$\gamma - g$		
		$\gamma - 1$	a'_1	$\gamma - 1$		
$g \leq 1$	$D = \gamma + g$	$g(\gamma + 1)$	b'_0	$g(\gamma + 1)$	$D = g\gamma + 1$	$g \leq 1$
		$g(\gamma - 1)$	b'_1	$g(\gamma - 1)$		
		$\gamma - g$	a'_1	$g\gamma - 1$		

Setting up the gain of the second-order filter is slightly trickier than for the first-order case, but once one works carefully through the definitions, the filter design equations are straightforward.

Second Order Shelf $(c_n = c'_n/D)$	Definitions: $G = \begin{cases} \dfrac{g\sqrt{2}}{2} & g > 2 \\ \sqrt{g} & 0.5 \le g \le 2 \\ g\sqrt{2} & g < 0.5 \end{cases}$ $g_d = \sqrt[4]{\dfrac{G^2-1}{g^2-G^2}} \quad g_n = g_d\sqrt{g} \quad \gamma = \tan\left(\pi f_c / F_s\right)$		
Low		High	
$D = g_d^2\gamma^2 + \sqrt{2}g_d\gamma + 1$		$D = \gamma^2 + \sqrt{2}g_d\gamma + g_d^2$	
$g_n^2\gamma^2 + \sqrt{2}g_n\gamma + 1$	b'_0	$\gamma^2 + \sqrt{2}g_n\gamma + g_n^2$	
$2(g_n^2\gamma^2 - 1)$	b'_1	$2(\gamma^2 - g_n^2)$	
$g_n^2\gamma^2 - \sqrt{2}g_n\gamma + 1$	b'_2	$\gamma^2 - \sqrt{2}g_n\gamma + g_n^2$	
$2(g_d^2\gamma^2 - 1)$	a'_1	$2(\gamma^2 - g_d^2)$	
$g_d^2\gamma^2 - \sqrt{2}g_d\gamma + 1$	a'_2	$\gamma^2 - \sqrt{2}g_d\gamma + g_d^2$	

Examples: F_s = 96 kHz: **1)** 1st-order Low, f_c = 300 Hz, g = 3; **2)** 1st-order Low, f_c = 300 Hz, g = 0.3333; **3)** 1st-order High, f_c = 8 kHz, g = 4; **4)** 1st-order High, f_c = 8 kHz, g = 0.25; **5)** 2nd-order Low, f_c = 100 Hz, g = 2; **6)** 2nd-order Low, f_c = 100 Hz, g = 0.5; **7)** 2nd-order High, f_c = 12 kHz, g = 3; **8)** 2nd-order High, f_c = 12 kHz, g = 0.3333. The coefficients for the first and last cases are shown in **Table 29**.

1 - Low f_c=300 g=3	$b_0 = 1.019444681090047 \quad b_2 = -0.961110637819907$ $a_1 = -0.980555318909954$
2 - High f_c=12k g=0.33	$b_0 = 0.425305792984813 \quad b_1 = -0.376798622612424$ $b_2 = 0.133506359063216$ $a_1 = -1.312549752924641 \quad a_2 = 0.494563282360246$

Table 29 Coefficients for Shelf Filters

The frequency response plots are shown in **Figure 109**. Notice that, as expected, the first-order filters do not transition as quickly as the second-order filters.

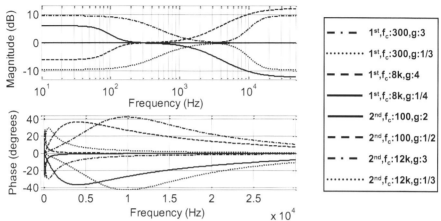

Figure 109 Frequency Responses of Shelf Filters

2.3.11 Troubleshooting the Filter Formulas

As you have seen above, this book contains many filter design equations. It also contains many examples for you to try so you can make sure you've understood the equations properly. If you have any trouble getting a filter to work out, go back to the relevant examples and make sure you can get the same values that I did as a first debugging step.

2.3.12 Fun with Poles and Zeros

So far we have designed filters that others thought up. These are very useful and can be configured as needed for most situations. However, filter design allows even more flexibility.

Think for a moment about the meaning of poles and zeros. If we write out a transfer function in factored analysis form, in terms of its poles and zeros, it is clear how it will behave. We will use a second-order example, but this is just as true for any order. The transfer function can be written as follows:

$$H(z) = \frac{b_0 + b_1 z^{-1} + b_2 z^{-2}}{1 + a_1 z^{-1} + a_2 z^{-2}} = \frac{b_0 z^2 + b_1 z + b_2}{z^2 + a_1 z + a_2} = \frac{k(z - z_1)(z - z_2)}{(z - p_1)(z - p_2)}$$

where k is a scale factor, z_1 and z_2 are the zeros, and p_1 and p_2 are the poles of the filter. Recall that the zeros and poles must either be real, or be complex conjugate pairs. Recall also that poles must not have magnitudes of one or larger, since that leads to instability.

Let us begin our analysis with a single zero and work from there. If there is only one zero, clearly it must be real. Recall also that zeros are not restricted in magnitude as are poles. The following transfer function contains only a single zero: $H(z) = \dfrac{k(z - z_1)}{1}$

Recall that the angles of poles and zeros relate to their active frequencies. A single pole or zero must lie on the real axis in the z-plane, and therefore can only affect frequencies near 0 Hz or $F_s/2$ Hz.

Figure 110 shows the effect of placing zeros at 0 Hz. Note that this is plotted on a linear frequency axis, since this makes the effect clearer. The gain factor in the equation is adjusted to always achieve 0 dB at $F_s/2$. The plot shows the effects of zeros of 0.1 (solid line, $z_1 = 0.1$), 0.5 (dashed line, $z_1 = 0.5$) and 1 (dash-dotted line, $z_1 = 1$). At lower right the figure also includes these zeros plotted in the z-plane. Note that the larger the magnitude of the zero, the greater the effect. Furthermore, when the zero is actually 1, the equation goes identically to 0, meaning the dash-dotted line in the figure is heading toward -∞ ($20 \cdot \log_{10}(0)$).

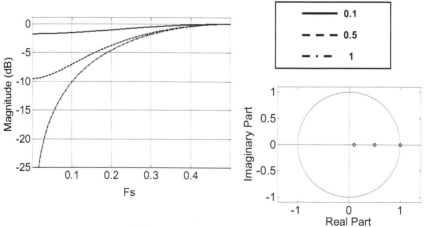

Figure 110 Zeros at 0 Hz

Now suppose we place these zeros at $F_s/2$ instead. Note that the same thing happens, although now at the opposite end of the spectrum (z_1 = -0.1, -0.5, -1) as shown in **Figure 111**.

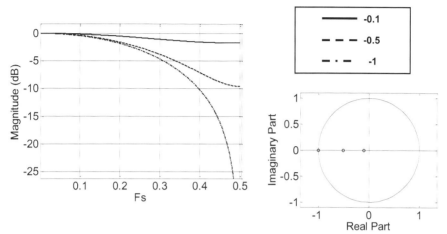

Figure 111 Zeros at $F_s/2$ Hz

To understand what happens when we place zeros between 0 and $Fs/2$, we must consider complex zeros, and, therefore, we must place two at a time since they must occur in complex conjugate pairs.

We will use the following equation $H(z) = \dfrac{k(z-z_1)(z-z_1^*)}{1}$

where * stands for complex conjugate. For our example, lets us try zeros at $F_s/8$, $F_s/4$ and $3F_s/8$. Recalling that 0 Hz to $F_s/2$ Hz maps to 0 to π, these correspond to $\pi/4$, $\pi/2$ and $3\pi/4$ respectively. Then, to also see the effects of different magnitudes, let us put magnitudes 1 at either end, and then magnitudes of 0.5 and 0.9 in the center of the band. The following calculation is for the first case.

$$1e^{\frac{\pi i}{4}} \to \frac{\sqrt{2}}{2} + i\frac{\sqrt{2}}{2}$$

Since we must have complex conjugate pairs,

$$z_1 = \frac{\sqrt{2}}{2} + i\frac{\sqrt{2}}{2} \quad z_1^* = \frac{\sqrt{2}}{2} - i\frac{\sqrt{2}}{2}$$

Figure 112 shows the result, where the solid line shows the magnitude 1 zeros at $\pi/4$, the dashed line shows the magnitude 1 zeros at $3\pi/4$, the dash-dotted line shows the magnitude 0.5 zeros at $\pi/2$ and the dotted line shows the magnitude 0.9 zeros, also at $\pi/2$.

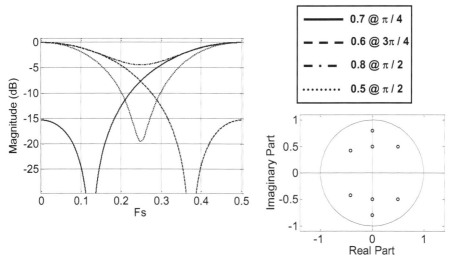

Figure 112 Zeros Across the Band

Now that we have considered a variety of zeros let us next try poles. **Figure 113** shows the effect of a pole of magnitude 0.6 at 0 (solid), and one of magnitude 0.3 at $F_s/2$ (dashed).

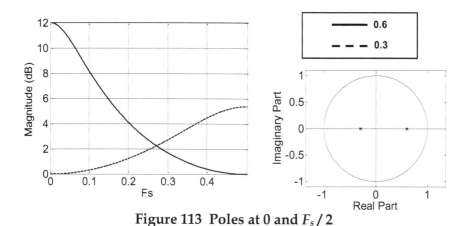

Figure 113 Poles at 0 and $F_s/2$

Chapter 3 IIR Filters

Figure 114 shows the effects of magnitude 0.7 poles at $\pi/4$ (solid), magnitude 0.6 poles at $3\pi/4$ (dashed), magnitude 0.8 poles at $\pi/2$ (dash-dotted) and magnitude 0.5 poles also at $\pi/2$ (dotted).

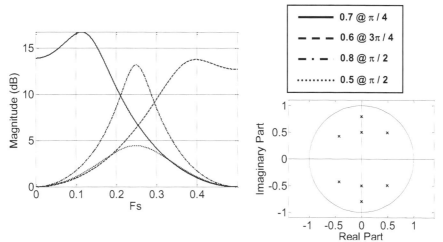

Figure 114 Poles Across the Band

Now, of course the greatest flexibility comes in combining poles and zeros together in a variety of patterns, using poles to push things up, and zeros to push things down. To better understand this, let us consider the case of a pair of fixed magnitude 0.5 poles at $\pi/2$, and what happens as we move some zeros in their vicinity. **Figure 115** shows the effect of this fixed set of poles along with a pair of 0.4 magnitude zeros at $\pi/2-0.1$ (solid), a pair of 0.6 magnitude zeros at $\pi/2+0.1$ (dashed), a pair of 0.7 magnitude zeros at $\pi/2+0.2$ (dash-dotted), and a pair of 0.8 magnitude zeros at $\pi/2-0.2$ (dotted). Clearly the flexibility is endless, and there is no reason why one must use only one pair each of poles and zeros.

2.3.13 After Note

Although most of the common IIR filter types are represented here, one filter you might have expected to see is missing: the Elliptical or Cauer filter. Unfortunately, this filter, which requires inversion of the Jacobi Elliptic Functions, does not lend itself to a simple formula. In many cases, one of the filters presented should be a fine substitute. If an Elliptical filter is necessary, it may be designed with many excellent software packages.

Digital Filters for Everyone

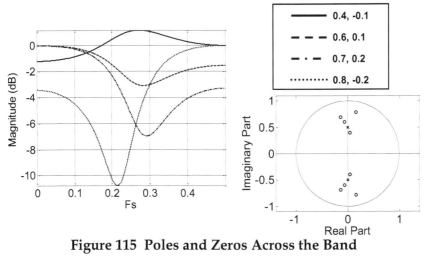

Figure 115 Poles and Zeros Across the Band

Chapter 4: Finite Impulse Response Filters

In a world of inexpensive processing resources, FIR filters enjoy significant popularity, which is not completely undeserved. These filters are inherently stable and, as will be shown shortly, it is simple to create linear phase FIR filters.

There are various references noted at the end of the book. **Reference 15** was particularly useful when writing this chapter.

Typically FIR filters are comparatively costly to implement, often requiring many hundreds or even thousands of delays and multiplies. However, the **Implementation, Tips and Tricks** chapter provides tips on making FIR filters implementation more cost effective. The reader should also not neglect the possibility of IIR filters. Finally, the **Advanced Topics** section also includes a method of designing linear phase IIR filters, though these are not necessarily less expensive to implement than efficiently-implemented FIR filters.

2.4.1 Linear Phase FIR Filters

While it is entirely possible to design non-linear phase FIR filters, the popularity of this filter stems largely from the inherent phase linearity of one important sub-class. Once this advantage is given up, many applications would probably benefit from the efficiency of IIR filters. Therefore, it is quite common to think of FIR filters as being inherently linear phase, though this is not quite accurate.

The requirement for linear phase FIR filters is that the coefficients be conjugate symmetric about the center. That is, if the coefficients are complex, those on the right of center should be complex conjugates of those on the left. In this book we are only interested in real coefficients, so simple symmetry will suffice, though it can actually be either actual- or anti-symmetry, as seen in our examples below.

2.4.2 Designing FIR Filters: Fourier Series Method

Readers of previous sections know that the focus is on simple design equations for a wide variety of digital filters. There are numerous methods available for designing FIR filters, and some of them, such as the popular Remez algorithm, are iterative and not suited to a simple formula. For those designs, a software solution may be appropriate. For others, simple design equations suffice while also giving the user insight into what is taking place within the design.

For some of these filters, you will see that the designs are slightly different when dealing with even or odd filter orders. In general, both sets of equations are given.

2.4.2.1 Lowpass

The following equations represent one method of computing coefficients for linear phase FIR filters.

Fourier Series Method Lowpass	Definition: $\gamma = \dfrac{2\pi f_c}{F_s}$
N Even	**N Odd**
$n = 0..\dfrac{N}{2}-1$	$n = 0..\dfrac{N-3}{2}$
(*m* fractional)	(*m* integral)
$h(n) = \dfrac{\sin(m\gamma)}{m\pi} \quad m = n - \dfrac{N-1}{2}$	
$h\left(\dfrac{N}{2}+k\right) = h\left(\dfrac{N}{2}-k-1\right)$ $k = 0..\dfrac{N}{2}-1$	$h\left(\dfrac{N-1}{2}\right) = \dfrac{\gamma}{\pi} = \dfrac{2f_c}{F_s}$ $h\left(\dfrac{N+2k+1}{2}\right) = h\left(\dfrac{N-2k-3}{2}\right)$ $k = 0..\dfrac{N-3}{2}$

Note that the **Windowing Functions** section below provides methods of modifying the coefficients in beneficial ways. For many designs one may choose to apply the equations from this section along with a window from that section, as is discussed below.

Note that N is the number of coefficients, and that, since one coefficient is always applied to the current sample, and since filter order is the number of delays, the filter order is N-1. In addition, notice that the equations for *m* and for *h* are the same for either case, but that m is fractional for the case where N is even, and integral for the case where N is odd. Furthermore, only in the case of N odd, there is a special consideration for the unique center coefficient. In the case where N is even, the two center coefficients are equal. Finally, although the last row of equations looks a little messy, the right half of the coefficients are just a mirror image of the left half, though for N odd one must be careful of the unique center value. Let us first try a couple of simple examples:

Example 1: $N = 6$; $f_c = 250$ Hz; $F_s = 10$ kHz

$$\gamma = \frac{2\pi f_c}{F_s} = \frac{2\pi 250}{10,000} = \frac{\pi}{20}$$

$$n = 0..\frac{6}{2} - 1 \rightarrow n = 0..2$$

$$h(n) = \frac{\sin(m\gamma)}{m\pi} = \frac{\sin\left(\left[n - \frac{N-1}{2}\right]\gamma\right)}{\left[n - \frac{N-1}{2}\right]\pi}$$

$$h(0) = \frac{\sin\left(\frac{-5\gamma}{2}\right)}{\frac{-5\pi}{2}} = \frac{-2\sin\left(\frac{-\pi}{8}\right)}{5\pi} = \frac{2\sin\left(\frac{\pi}{8}\right)}{5\pi} = 0.048724767920222$$

$$h(1) = \frac{\sin\left(\frac{-3\gamma}{2}\right)}{\frac{3\pi}{2}} = \frac{-2\sin\left(\frac{-3\pi}{40}\right)}{3\pi} = \frac{2\sin\left(\frac{3\pi}{40}\right)}{3\pi} = 0.049538644799405$$

$$h(2) = \frac{\sin\left(\frac{-\gamma}{2}\right)}{\frac{-\pi}{2}} = \frac{-2\sin\left(\frac{-\pi}{40}\right)}{\pi} = \frac{2\sin\left(\frac{\pi}{40}\right)}{\pi} = 0.049948611662427$$

Note that sine is an odd function: $\sin(-x) = -\sin(x)$. This is how the signs were cancelled in the right hand equations above. The remaining coefficients are just a reflection of these:

$$h\left(\frac{N}{2} + k\right) = h\left(\frac{N}{2} - k - 1\right) \rightarrow h\left(\frac{6}{2} + k\right) = h\left(\frac{6}{2} - k - 1\right)$$

$$h(3+0) = h(3-1) \to h(3) = h(2) = 0.049948611662427$$
$$h(3+1) = h(3-1-1) \to h(4) = h(1) = 0.049538644799405$$
$$h(3+2) = h(3-2-1) \to h(5) = h(0) = 0.048724767920222$$

$$h = \begin{bmatrix} 0.048724767920222, 0.049538644799405, \\ 0.049948611662427, 0.049948611662427, \\ 0.049538644799405, 0.048724767920222 \end{bmatrix}$$

Now let us repeat this example for N odd:

Example 2: $N = 7$; $f_c = 250$ Hz; $F_s = 10$ kHz

$$\gamma = \frac{2\pi f_c}{F_s} = \frac{2\pi 250}{10,000} = \frac{\pi}{20}$$

$$n = 0..\frac{7-3}{2} \to n = 0..2$$

$$h(n) = \frac{\sin(m\gamma)}{m\pi} = \frac{\sin\left(\left[n - \frac{N-1}{2}\right]\gamma\right)}{\left[n - \frac{N-1}{2}\right]\pi}$$

$$h(0) = \frac{\sin(-3\gamma)}{-3\pi} = \frac{\sin\left(\frac{3\pi}{20}\right)}{3\pi} = 0.048169888100206$$

$$h(1) = \frac{\sin(-2\gamma)}{-2\pi} = \frac{\sin\left(\frac{\pi}{10}\right)}{2\pi} = 0.049181582154173$$

$$h(2) = \frac{\sin(-\gamma)}{-\pi} = \frac{\sin\left(\frac{\pi}{20}\right)}{\pi} = 0.049794636762178$$

Now, since N is odd, we compute the unique center coefficient:

$$h\left(\frac{N-1}{2}\right) = \frac{\gamma}{\pi} = \frac{2f_c}{F_s} \to h(3) = \frac{500}{10,000} = \frac{1}{20} = 0.05$$

Once again the right side is just a reflection of the left side, not including the unique center:

$$h\left(\frac{N+2k+1}{2}\right) = h\left(\frac{N-2k-3}{2}\right) \rightarrow h\left(\frac{7+2k+1}{2}\right) = h\left(\frac{7-2k-3}{2}\right)$$

$$h\left(\frac{7+1}{2}\right) = h\left(\frac{7-3}{2}\right) \rightarrow h(4) = h(2) = 0.049794636762178$$

$$h\left(\frac{7+2+1}{2}\right) = h\left(\frac{7-2-3}{2}\right) \rightarrow h(5) = h(1) = 0.049181582154173$$

$$h\left(\frac{7+4+1}{2}\right) = h\left(\frac{7-4-3}{2}\right) \rightarrow h(6) = h(0) = 0.048169888100206$$

$$h = \begin{bmatrix} 0.048169888100206, 0.049181582154173, 0.049794636762178, 0.05, \\ 0.049794636762178, 0.049181582154173, 0.048169888100206 \end{bmatrix}$$

Generally speaking these equations result in filters with passbands scaled near 0 dB. However, when the filter length is very short, and particularly for lower cutoff frequencies, it is normal for the passband to be scaled lower as shown in **Figure 116**.

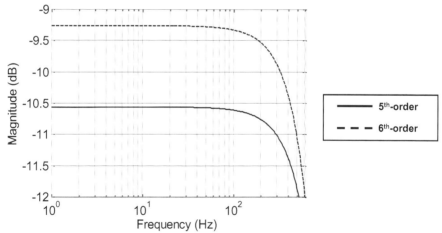

Figure 116 Original Passband Scaling

The **Tips and Tricks** section discusses scaling in more detail, but for an FIR filter, it is possible to set the response at 0 Hz to 1 by making the sum of the coefficients equal to 1. This is easily accomplished by dividing each of the coefficients by the original sum of coefficients. It is also possible to just multiply by the linear gain that will change the magnitude by the amount desired. For example, if you wish to increase it by 6 dB, multiply by 2.

For the fifth-order filter ($N=6$):

$$h = \begin{bmatrix} 0.048724767920222, 0.049538644799405, \\ 0.049948611662427, 0.049948611662427, \\ 0.049538644799405, 0.048724767920222 \end{bmatrix}$$

$$s = \sum_{n=0}^{5} h(n) = 0.296424048764106$$

$$h' = \frac{h}{s} = \begin{bmatrix} 0.164375218958691, 0.167120869598631, \\ 0.168503911442678, 0.168503911442678, \\ 0.167120869598631, 0.164375218958691 \end{bmatrix}$$

For the sixth-order filter ($N=7$):

$$h = \begin{bmatrix} 0.048169888100206, 0.049181582154173, 0.049794636762178, 0.05, \\ 0.049794636762178, 0.049181582154173, 0.048169888100206 \end{bmatrix}$$

$$s = \sum_{n=0}^{6} h(n) = 0.344292214033114$$

$$h' = \frac{h}{s} = \begin{bmatrix} 0.139909896700635, 0.142848371672567, 0.144628994594077, \\ 0.145225474065443, 0.144628994594077, 0.142848371672567, \\ 0.139909896700635 \end{bmatrix}$$

The resulting filter responses are shown in **Figure 117**. Notice that the sixth-order filter achieves a slightly lower passband attenuation (at the top of the ripple) and cuts off somewhat faster. This is typical behavior and the filters will perform even better for higher orders. (Fifth- or sixth-order filters are very short; FIR filters often have hundreds or even thousands of coefficients.)

Figure 118 and **Figure 119** show additional filter analysis for these two filters. Since the filters are so short, the impulse responses show little of the character that we will soon see in longer filters. Recall that, for an FIR filter, the impulse response is a replica of the filter coefficients, so the plots show the actual coefficients.

Chapter 4 FIR Filters

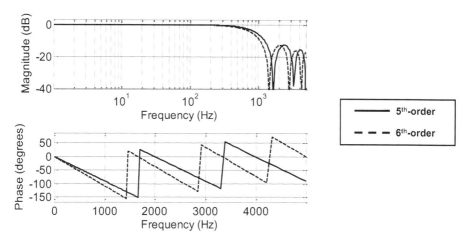

Figure 117 Frequency Responses of Rescaled Filters

Notice that the zeros are on the unit circle; since they are FIR filters, their poles are at the origin. These filters have exactly linear phase so their group delays are constant across the band. In addition, the group delay for the fifth-order case ($N = 6$) is 2.5, and that for the sixth-order case it is 3. In cases where it is useful to compensate for this delay, the exact integer number is preferable since it can be compensated by delaying another path by that many samples.

These small filters were used to show the calculation details. Now let us try some more typical, longer examples. For the sake of comparison, let us try the same parameters other than N, which will be 150 and 175 for the two new cases.

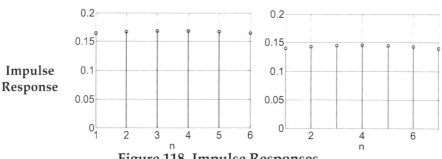

Figure 118 Impulse Responses

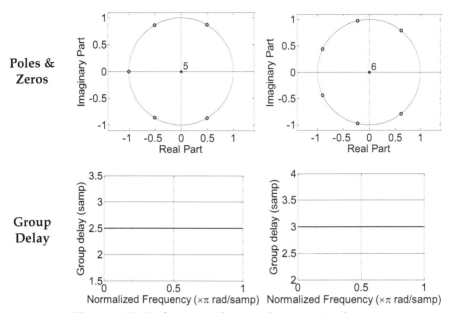

Figure 119 Pole Zero Plots and Group Delays

For N = 150, the coefficients near the center are as follows:

$h(71) = 0.047518988190265 \quad h(72) = 0.048724767920222$

$h(73) = 0.049538644799405 \quad h(74) = 0.049948611662427$

As before, these are mirrored after this point: $h(75) = h(74)$ and so forth. For N = 175, the coefficients near the center are as follows:

$h(85) = 0.048169888100206 \quad h(85) = 0.049181582154173$

$h(86) = 0.049794636762178 \quad h(87) = 0.050000000000000$

These are mirrored after this point, except for the center value: $h(88) = h(86)$ and so on.

The frequency responses of these two filters are plotted in **Figure 120**. Note that the magnitude response achieves a lower value in the passband as compared to the lower-order cases of **Figure 117**. Notice also that the passband is essentially 0 dB without any rescaling. This is typical for longer filters using this technique. Also note that there is a little ringing (overshoot) in the passband. We will discuss this more in the **Window Functions** section below.

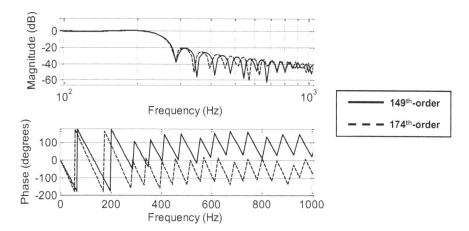

Figure 120 Lowpass Frequency Responses

Additional analyses for these two filters are shown in **Figure 121** and **Figure 122** below. Notice the distinctive shape of the impulse responses. This sinc function (sin(x)/x) shape is typical, when there are enough samples to produce it. Of course that is to be expected given that is the function used to generate the coefficients. This shape, however, will be modified somewhat to some benefit in the **Window Functions** section below.

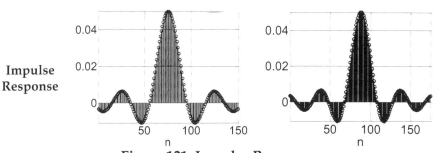

Figure 121 Impulse Reponses

Poles & Zeros

Group Delay

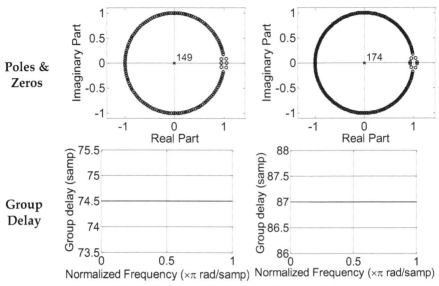

Figure 122 Pole Zero Plots and Group Delays

2.4.2.2 Highpass

Similar to the lowpass case, the highpass coefficients can be computed using the equations below.

Fourier Series Method Highpass		Definition: $\gamma = \dfrac{2\pi f_c}{F_s}$	
N Even		**N Odd**	
$n = 0..\dfrac{N}{2}-1$		$n = 0..\dfrac{N-3}{2}$	
(m fractional)	$m = n - \dfrac{N-1}{2}$	(m integral)	
$h(n) = \dfrac{\cos(m\gamma)}{m\pi}$		$h(n) = \dfrac{-\sin(m\gamma)}{m\pi}$	
$h\left(\dfrac{N}{2}+k\right) = -h\left(\dfrac{N}{2}-k-1\right)$ $k = 0..\dfrac{N}{2}-1$		$h\left(\dfrac{N-1}{2}\right) = 1 - \dfrac{\gamma}{\pi} = 1 - \dfrac{2f_c}{F_s}$ $h\left(\dfrac{N+2k+1}{2}\right) = h\left(\dfrac{N-2k-3}{2}\right)$ $k = 0..\dfrac{N-3}{2}$	

As examples, let us consider highpass filters with $N = 112$ and 133, with cutoff frequencies of 3 kHz and sample rates of 15 kHz. The frequency responses of these filters are shown in **Figure 123**. Note that the higher order filter achieves a greater stopband attenuation as expected. However, also note that the even-ordered filter has a flat response at 0 Hz while the odd-ordered filter is continuing on a downward trend. This is a characteristic of highpass filters designed in this fashion. If the additional attenuation at very low frequencies is important, an odd-ordered filter might be preferred, and a higher odd order will still achieve additional stopband attenuation.

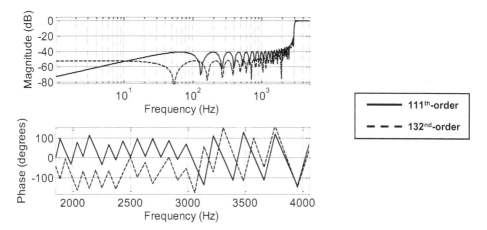

Figure 123 Highpass Frequency Responses

It is also important to note that these filters are still linear phase, and, therefore, have constant group delay. The 132nd-order filter has a group delay of 66 samples, making delay compensation easy, if that is important. The 111st-order filter has a group delay of 55.5 samples, meaning compensation, if required, is non-trivial. Therefore, each filter has its merits and the designer can choose the characteristics that best suit the situation.

The impulses responses for each case are shown in **Figure 124** below, with the 111st-order filter on the left, and the 132nd-order filter on the right. The second row is a zoom around the center. Note that the odd-order (even number of coefficients, 112) filter is actually anti-symmetric about the center, which still produces linear phase. The even-order filter (odd number of coefficients, 133) is symmetric, with a center coefficient of opposite sign as compared to its two neighbors. These are characteristics of these filters.

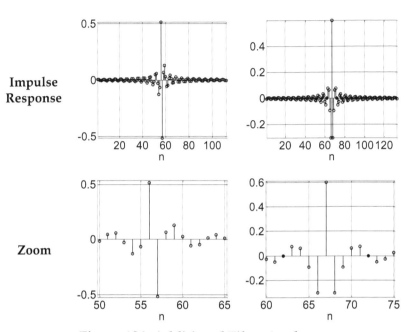

Figure 124 Additional Filter Analyses

2.4.2.3 Bandpass

The design equations for FIR bandpass filters are shown below.

As examples, let us design two bandpass filters with a passband between 1 kHz and 2 kHz for a sample rate of 20 kHz. Let us try one filter with $N = 150$, and the other with $N = 199$. The associated frequency responses are plotted in **Figure 125**.

As always, the group delays for these filters are half the filter order or $(N-1)/2$: 74.5 and 99 respectively. The impulse responses are shown in **Figure 126**.

Chapter 4 FIR Filters

Fourier Series Method Bandpass	Definitions: $\gamma_1 = \dfrac{2\pi f_{c1}}{F_s} \quad \gamma_2 = \dfrac{2\pi f_{c2}}{F_s}$	
N Even	**N Odd**	
$n = 0 .. \dfrac{N}{2} - 1$	$n = 0 .. \dfrac{N-3}{2}$	
(m fractional)	$m = n - \dfrac{N-1}{2}$	(m integral)
$h(n) = \dfrac{\sin(m\gamma_2) - \sin(m\gamma_1)}{m\pi}$		
$h\left(\dfrac{N}{2}+k\right) = h\left(\dfrac{N}{2}-k-1\right)$ $k = 0 .. \dfrac{N}{2} - 1$	$h\left(\dfrac{N-1}{2}\right) = \dfrac{\gamma_2 - \gamma_1}{\pi} = \dfrac{2(f_{c2} - f_{c1})}{F_s}$	
	$h\left(\dfrac{N+2k+1}{2}\right) = h\left(\dfrac{N-2k-3}{2}\right)$ $k = 0 .. \dfrac{N-3}{2}$	

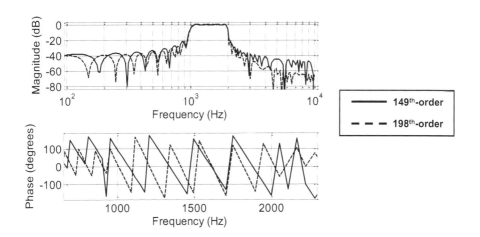

Figure 125 Bandpass Frequency Responses

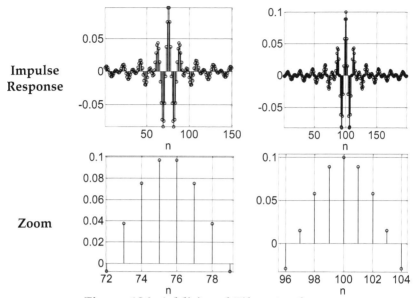

Figure 126 Additional Filter Analyses

2.4.2.4 Bandstop

The design equations for FIR bandstop filters are shown below. These filters must have even order, or odd N.

Fourier Series Method Bandstop	Definitions: $\gamma_1 = \dfrac{2\pi f_{c1}}{F_s}$ $\gamma_2 = \dfrac{2\pi f_{c2}}{F_s}$	
N Even	**N Odd**	
Not Defined	$n = 0 .. \dfrac{N-3}{2}$	
	$m = n - \dfrac{N-1}{2}$	(m integral)
	$h(n) = \dfrac{\sin(m\gamma_1) - \sin(m\gamma_2)}{m\pi}$	
	$h\left(\dfrac{N-1}{2}\right) = 1 + \dfrac{\gamma_1 - \gamma_2}{\pi} = 1 + \dfrac{2(f_{c1} - f_{c2})}{F_s}$	
	$h\left(\dfrac{N+2k+1}{2}\right) = h\left(\dfrac{N-2k-3}{2}\right)$	
	$k = 0 .. \dfrac{N-3}{2}$	

As an example, let's use the same parameters as above and design a bandstop filter with a stopband between 1 kHz and 2 kHz for a sample rate of 20 kHz. Since N must be odd, we shall do only the case of $N = 199$. The frequency response is plotted in **Figure 127**.

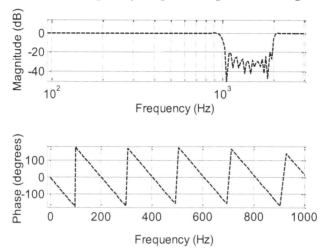

Figure 127 Bandstop Frequency Response

The group delay is 99 and the impulse response is shown in **Figure 128** with a zoom in **Figure 129**. Comparing the design equations for the bandstop filter with those for the bandpass filter, it is clear that the center coefficient of the bandstop filter is one minus that for the bandpass; the other coefficients have the same magnitudes and the reverse sign, for the same order filter. This relationship is also apparent by comparing the impulse responses of the two cases.

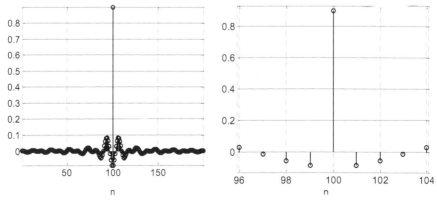

Figure 128 Impulse Response   **Figure 129 Imp. Response Zoom**

2.4.2.5 Window Functions and Variations

Though we have not focused on this issue, the filters in the previous section all exhibit ringing in the passband. For example, note the low-frequency passband of the bandstop filter of **Figure 127**, a zoom of which is plotted in **Figure 130**.

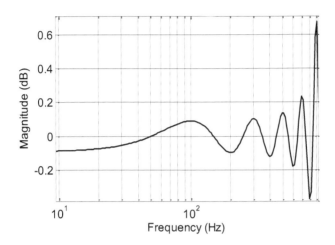

Figure 130 Bandstop Frequency Response Ripple

Called Gibbs Phenomenon [**Reference 6** page 251], this ringing is typically undesirable but easy to address by the application of a window function, as discussed in this section.

There exist numerous windows [**Reference 6** page 258] with various properties. The application of no window at all, as we did above, is typically referred to as the *Rectangular Window*. For the purposes of this book, we have chosen to focus on three additional window functions due to their combination of simplicity and utility: Cosine, Hamming, and von Hann (sometimes called Hann or Hanning).

N Even	N Odd
$n = -\dfrac{N-1}{2} .. \dfrac{N}{2}$	$n = -\dfrac{N-1}{2} .. \dfrac{N-1}{2}$
$h(n) = \cos\left(\dfrac{n\pi}{N+1}\right)$	
Cosine Window	

As examples we shall show a few values of the cases, $N = 10$, $N = 23$:

$W_{C10} = 0.281732556841430, \ldots 0.755749574354258, \ldots 0.989821441880933, \ldots$
$W_{C23} = 0.130526192220052, \ldots 0.500000000000000, \ldots 1.000000000000000, \ldots$

N Even	N Odd
$n = -\dfrac{N}{2} \ldots \dfrac{N}{2} - 1$	$n = -\dfrac{N-1}{2} \ldots \dfrac{N-1}{2}$
$h(n) = 0.54 + 0.46 \cos\left(\dfrac{\pi(2n+1)}{N-1}\right)$	$h(n) = 0.54 + 0.46 \cos\left(\dfrac{\pi 2n}{N-1}\right)$
Hamming Window	

$W_{H10} = 0.080000000000000, \ldots 0.460121838273212, \ldots 0.972258605561518, \ldots$
$W_{H23} = 0.080000000000000, \ldots 0.238764062385169, \ldots 1.000000000000000, \ldots$

N Even	N Odd
$n = -\dfrac{N}{2} \ldots \dfrac{N}{2} - 1$	$n = -\dfrac{N-1}{2} \ldots \dfrac{N-1}{2}$
$h(n) = 0.5 + 0.5 \cos\left(\dfrac{\pi(2n+1)}{N-1}\right)$	$h(n) = 0.5 + 0.5 \cos\left(\dfrac{\pi 2n}{N-1}\right)$
von Hann Window	

$W_{V10} = 0, 0.116977778440511, \ldots 0.750000000000000, 0.969846310392954, \ldots$
$W_{V23} = 0, 0.020253513192751, \ldots 0.707707506500943, 1.000000000000000, \ldots$

Now, to reduce the effect of the Gibbs phenomenon, one merely computes the FIR filter coefficients as above, then computes the chosen window of the appropriate size, and multiplies the filter coefficients by the window, on a point by point basis. Since all of these windows are symmetrical, the filter symmetry is not changed when the window is applied. Let us try a few examples.

Figure 131 shows the three windows sized at 199 samples. Table 30 shows three coefficients each from the four different varieties of window: Rectangular (original), Cosine, Hamming and von Hann. These correspond to coefficients 50 – 52, where coefficients are numbered beginning at 0.

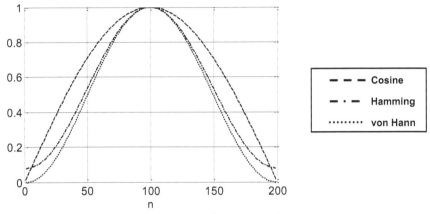

Figure 131 Window Functions

The upper pane of **Figure 132** is a repeat of **Figure 130** but this time with the additional responses arising from applying the three windows above to the filter coefficients. On this scale, the windowed responses are all very flat as compared to the original (solid line, marked as *Rectangular* in the legend). When the original filter response is suppressed and the scale zoomed, as in the lower pane (note that the highest value in this plot is $10 \times 10^{-3} = 0.01$ dB), it is clear that the windowed responses also have ripple, though much less as compared to the non-windowed (rectangular) response.

Rectangular	0.005825735123797	0.010204761421125	0.011920187209713
Cosine	0.004183613596202	0.007438950926320	0.008816541118270
Hamming	0.003188415182132	0.005733929732920	0.006871451281408
Von Hann	0.002959083013291	0.005345161760033	0.006432430765904

Table 30 Three Coefficients

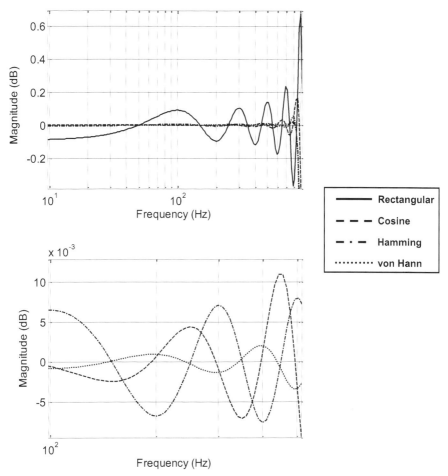

Figure 132 Windowed Bandstop Frequency Response

The windows also change the transition and stopband performance of the filter, as seen in **Figure 133**. The non-windowed (rectangular) filter does transition slightly quicker, but then has significant stopband ripple, while the various windowed versions achieve greater stopband attenuation and less stopband ripple. Note also that applying the windows does not change the implementation cost of the filter, since it only modifies the filter coefficients. Therefore, generally speaking, it is worth the extra effort to apply a window when creating FIR filters using the methods of this section.

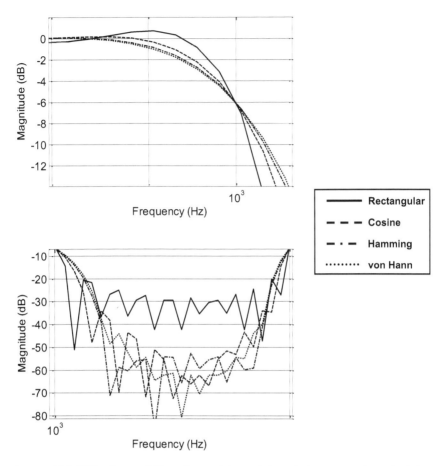

Figure 133 Windowed Bandstop Transition (U) & Stopbands (L)

2.4.3 Designing FIR Filters: Frequency Sampling Method

Another straightforward FIR filter design technique is the *frequency sampling method*. Using this method, the desired frequency response is defined using a vector of length equal to half (rounding up for odd lengths) the desired number of filter taps (filter length N). This vector is then manipulated with one of the equations below to determine the filter coefficients.

As with the FIR filters designed above using the Fourier series method, those designed with the frequency sampling method will achieve higher performance with a higher number of coefficients. In the case of the frequency sampling method, the number of coefficients is also a consideration when choosing cutoff frequencies, because it relates to the resolution available to make changes to the filter's transfer function, as explained below.

Setting Up the Desired Response: The desired response vector is comprised of (linear, not dB) gains on a uniform grid of $N/2$ frequencies between 0 Hz and $F_s/2$. Let us try a couple of examples:

Case 1: Lowpass; $N = 16$; $F_s = 10$ kHz; $f_c = 2.5$ kHz

If there are to be N coefficients (taps), there will be $\left\|\frac{N}{2}-1\right\|$ (round to nearest integer) intervals between 0 and $F_s/2$. For our example, there are 16 coefficients, so there are 7 intervals and the frequency sample points fall at 0, 5000/7, 10000/7, 15000/7, 20000/7, 25000/7, 30000/7, 5000 Hz, since $F_s/2$ is 5000 Hz. We wish for everything below 2.5 kHz to be 1, and everything else to be 0. Following the equation: 15000/7 is 2.14 kHz and 20000/7 = 2.86 kHz. So we assign a gain of 1 to all frequencies between 0 and 15000/7, and assign all other frequencies 0, as shown in the table below. (Note that there are other considerations to get the cutoff frequency right. We will discuss that further below.)

F (kHz)	0	0.7	1.4	2.1	2.9	3.6	4.3	5
A	1	1	1	1	0	0	0	0

Case 2: Highpass; $N = 17$; $F_s = 23$ kHz; $f_c = 5$ kHz

In this case, if we round 17/2 - 1 we get 8. So we will have 8 intervals between 0 and 11.5 kHz: 0, 11.5k / 8, 23k/8 .. 11.5k. Since this is a highpass, the vector is comprised of zeros on the left side and ones on the right as shown below. Once again, we will discuss other options for optimizing filter performance below.

F (kHz)	0	1.4	2.9	4.3	5.8	7.2	8.6	10.1	11.5
A	0	0	0	0	1	1	1	1	1

Of course, we can construct vectors for bandpass and bandstop filters by placing ones in the desired passbands, and zeros in the desired stopbands.

Shown below are equations for creating filters from the vectors created above. Note the slight variation for odd and even filters.

Frequency Sampling Method

N Even	N Odd
$n = 0..\left\lVert\dfrac{N}{2}-1\right\rVert$	$M = \dfrac{N-1}{2}$
(M fractional)	(M integral)
$h(n) = \dfrac{A(0)+2\sum_{k=1}^{\lfloor M \rfloor} A(k)\cos\left(\dfrac{2\pi k(n-M)}{N}\right)}{N}$	
$h\left(\dfrac{N}{2}+k\right) = h\left(\dfrac{N}{2}-k-1\right)$	$h\left(\dfrac{N+2k+1}{2}\right) = h\left(\dfrac{N-2k-3}{2}\right)$
$k = 0..\dfrac{N}{2}-1$	$k = 0..\dfrac{N-3}{2}$

The double bars in the equation for n denote rounding (4.5 becomes 5), while the topless brackets around M in the summation denote the floor function, or truncation down to the next integer (4.5 becomes 4). Let us now use these equations to design the filters for which we set up vectors above.

Case 1 (from above): Lowpass; $N = 16$; $F_s = 10$ kHz; $f_c = 2.5$ kHz

F (kHz)	0	0.7	1.4	2.1	2.9	3.6	4.3	5
A	1	1	1	1	0	0	0	0

$$n = 0..\left\lVert\dfrac{16}{2}-1\right\rVert = 0..7 \qquad M = \dfrac{16-1}{2} = 7.5$$

$$h(n) = \dfrac{A(0)+2\sum_{k=1}^{7} A(k)\cos\left(\dfrac{2\pi k(n-7.5)}{16}\right)}{16}$$

$$h(0) = \dfrac{1+2\sum_{k=1}^{7} A(k)\cos\left(\dfrac{2\pi k(-7.5)}{16}\right)}{16}$$

$$= \dfrac{1+2\left(A(1)\cos\left(\dfrac{2\pi(-7.5)}{16}\right)+A(2)\cos\left(\dfrac{4\pi(-7.5)}{16}\right)+...+A(7)\cos\left(\dfrac{14\pi(-7.5)}{16}\right)\right)}{16}$$

$$= \frac{1 + 2\left(\cos\left(\frac{2\pi(-7.5)}{16}\right) + \cos\left(\frac{4\pi(-7.5)}{16}\right) + \cos\left(\frac{6\pi(-7.5)}{16}\right)\right)}{16}$$

$$h(0) = -0.048546920024311$$

$$\vdots$$

$$h(7) = \frac{1 + 2\sum_{k=1}^{7} A(k)\cos\left(\frac{2\pi k(7-7.5)}{16}\right)}{16}$$

$$= \frac{1 + 2\left(\cos\left(\frac{2\pi(-0.5)}{16}\right) + \cos\left(\frac{4\pi(-0.5)}{16}\right) + \cos\left(\frac{6\pi(-0.5)}{16}\right)\right)}{16}$$

$$h(7) = 0.404516803152133$$

Note that since the values of $A(k)$ are 0 for $k = 4..7$, the cosine term disappears for those cases. The frequency response for this filter is shown in **Figure 134** below. Directly beneath the overall response are zooms of the passband ripple and the cutoff response.

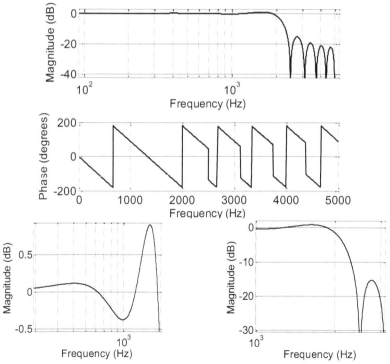

Figure 134 Frequency Response

Clearly there is an overshoot in the passband of nearly 1 dB, and an undershoot of around half that value. In addition, the filter hits a null at 2.5 kHz; typically it is desirable for the attenuation to be only a few dB at the cutoff frequency.

In the case of the Fourier series design method we used window functions to both soften the transition and mitigate the ripple. It would certainly be possible to apply window functions to the coefficients generated through the frequency sampling method, but other tools are also available; we can use values other than 0 or 1 in the desired response as shown below:

F (kHz)	0	0.7	1.4	2.1	2.9	3.6	4.3	5
A	1	1	1	0.9	0.5	0	0	0

Here, we have placed a value of 0.9 at 1.4 kHz to combat the ripple, and have also placed a value of 0.5 at 2.9 kHz to combat the high roll off at the cutoff frequency. The filter arising from this new vector is shown in **Figure 135** plotted against the original response.

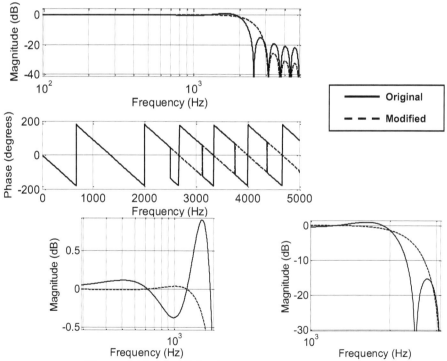

Figure 135 Modified Frequency Response

Note how the passband ripple has been reduced significantly while the filter has a much lower roll off at the cutoff frequency.

Using more coefficients can also help both of these problems in two ways: 1) by providing greater frequency resolution and thereby greater precision in selecting the desired inflections, and 2) the higher filter order generally allows more flexibility in filter response, as has been seen for other filters as well.

Figure 136 and **Figure 137** shows additional analyses of these filters.

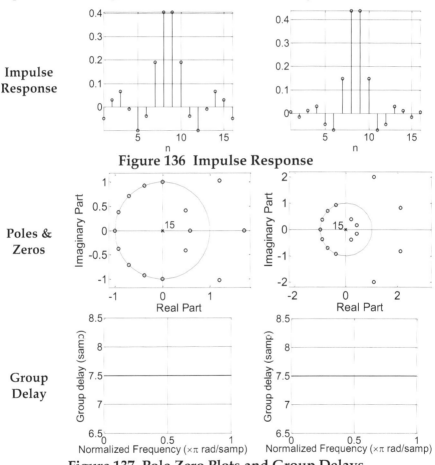

Figure 136 Impulse Response

Figure 137 Pole Zero Plots and Group Delays

Note from the impulse response plots, that the effect on the coefficients of modifying the desired response plots was similar to the taper that occurs when applying a window, as we did using the

Fourier Series Method. Notice also that the zero patterns are similar, except that the right half of the plane contains zeros of slightly higher magnitudes. Finally, notice that the group delays are identical; since both versions of the filter are completely linear phase, the group delays are exactly flat and valued at half the filter order.

Now let us design a filter based upon Case 2 from above.

Case 2: Highpass; $N = 17$; $F_s = 23$ kHz; $f_c = 5$ kHz

F (kHz)	0	1.4	2.9	4.3	5.8	7.2	8.6	10.1	11.5
A	0	0	0	0	1	1	1	1	1

We have plotted in **Figure 138** the original and an alternative filter that reduces passband ripple, the vector for which is shown below.

F (kHz)	0	1.4	2.9	4.3	5.8	7.2	8.6	10.1	11.5
A	0	0	0	0.25	0.75	1	1	1	1

Note that the alternative filter has a similar cutoff response, but then a significantly deeper stopband attenuation than the original.

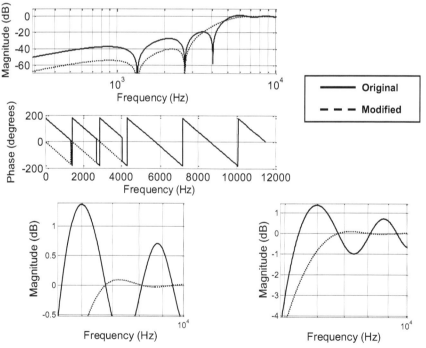

Figure 138 Case 2 Frequency Responses

Now, let us try a few more cases.

Case 3: Bandpass; $N = 171$; $F_s = 125$ kHz; $f_{cL} = 25$ kHz; $f_{cH} = 35$ kHz

In this case, rounding up from 171/2 - 1 gives 86 elements in the vector. Furthermore, there will be 85 intervals, meaning $f_{cL} = n_L F_s / (2 \times 85) = 25$ kHz, and $f_{cH} = n_H F_s / (2 \times 85) = 35$ kHz. From this, $n_L = 34$ and $n_H \approx 48$. Based on lessons learned regarding softening the transition band, let the vector, V, be all zeros except as follows: $V(33) = 0.25$, $V(34) = 0.75$, $V(35:47) = 1$, $V(48) = 0.75$, $V(49) = 0.25$. The resulting filter is plotted in **Figure 139**.

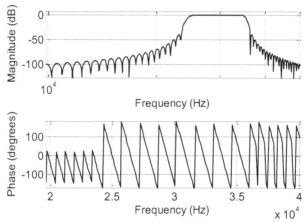

Figure 139 Bandpass Frequency Response

Finally, let us try a bandstop filter in Case 4.

Case 4: **Figure 140** shows the frequency response of a bandstop filter with the following characteristics: $N = 98$; $F_s = 100$ kHz; $f_{cL} = 10$ kHz; $f_{cH} = 20$ kHz. In this case, the vector is 49 samples long meaning there are 48 intervals: $f_{cL} = n_L F_s / (2 \times 48) = 10$ kHz, and $f_{cH} = n_H F_s / (2 \times 48) = 20$ kHz. From this, $n_L \approx 10$ and $n_H \approx 19$. For this bandstop filter, V is all ones with the following exceptions: $V(10) = 0.75$, $V(11) = 0.25$, $V(12:17) = 0$, $V(18) = 0.25$, $V(19) = 0.75$.

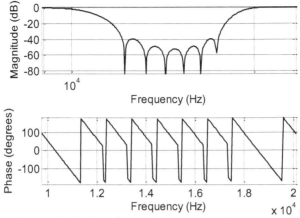
Figure 140 Bandstop Frequency Response

2.4.4 Fun with Zeros

In our IIR chapter, **Fun with Poles and Zeros** section, we discussed placing poles and zeros to create arbitrary filters. The zero placement part of that discussion is relevant here as well. In addition, let us consider further properties of zeros and of FIR filters.

To begin with, let us consider a complex conjugate pair of zeros. Often we think of these in rectangular coordinates with zeros of the form $a \pm ib$; but let us instead write out the equation in polar coordinates, where the zeros are of the form $re^{\pm i\theta}$:

$$(z - re^{i\theta})(z - re^{-i\theta})$$
$$z^2 - r(e^{i\theta} + e^{-i\theta})z + r^2$$

We recall from Euler's formulas that $\cos(\theta) = \dfrac{e^{i\theta} + e^{-i\theta}}{2}$. Therefore, the equation above becomes

$$z^2 - 2r\cos(\theta)z + r^2 \qquad \text{Equation 36}$$

Second-order FIR Filter in Polar Coordinates

This form is useful because it guides us in placing our zeros at particular angles on the unit circle, which in turn associate with particular frequencies.

In addition, recall that for linear phase filters we need symmetry of coefficients. This means that our second-order FIR filter, as described above, will be linear phase if we make $r = 1$ (putting the zeros on the unit circle). What is more, we will even be able to combine many such filters and still retain this symmetry, as we shall see shortly. Note that we can also scale the entire filter by an arbitrary value without upsetting the symmetry, and without changing the zero locations. This allows us to set the passband gain as desired. The topic of scaling is discussed further in the **Tips and Tricks** chapter.

As an example of the technique above, imagine that we wish to place zeros on the unit circle at 0.23π, 0.46π, 0.69π and 0.92π. We can do this by first creating four second-order sections using the equation above and then combining them to create a single eighth-order FIR filter.

$$z^2 - 2\cos(0.23\pi)z + 1 = z^2 - 1.5002z + 1$$
$$z^2 - 2\cos(0.46\pi)z + 1 = z^2 - 0.2507z + 1$$
$$z^2 - 2\cos(0.69\pi)z + 1 = z^2 + 1.1242z + 1$$
$$z^2 - 2\cos(0.92\pi)z + 1 = z^2 + 1.9372z + 1$$

To combine these we use an operation called convolution which is denoted by the symbol $\otimes$. In this case, it operates much like integer multiplication[1]. The coefficients from these equations form the vectors [1 -1.5002 1], [1 -0.2507 1], [1 1.1242 1] and [1 1.9372 1]. We shall combine the first two, then combine that combination with the next one, and so forth:

$$[1 \quad -1.5002 \quad 1] \otimes [1 \quad -0.2507 \quad 1] =$$

```
     1   -1.5002   1
+   -0.2507   0.3761   -0.2507
+             1        -1.5002   1
= 1  -1.7509   2.3761   -1.7509   1
```

In like manner,

$$[1 \quad -1.7509 \quad 2.3761 \quad -1.7509 \quad 1] \otimes [1 \quad 1.1242 \quad 1]$$
$$= 1.0000 \quad -0.6267 \quad 1.4078 \quad -0.8307 \quad 1.4078 \quad -0.6267 \quad 1.0000$$

---

1. Note that for convolution, we actually must flip one of the vectors before carrying out the multiplication. However, when the vectors are symmetric, the same result is achieved with or without the flip.

and

[1.0000 -0.6267 1.4078 -0.8307 1.4078 -0.6267 1.0000]⊗[1 1.9372 1]
= 1.0000 1.3104 1.1937 1.2697 1.2063 1.2697 1.1937 1.3104 1.0000

This last line is our new FIR filter. Note the symmetry. The frequency response of this filter is shown in **Figure 141**. The zeros we placed have caused nulls in the frequency response at exactly the specified angles (frequencies).

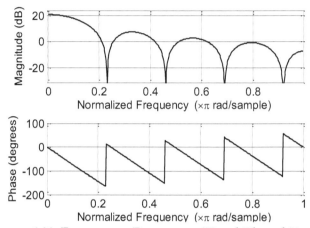

Figure 141 Frequency Response: Hand Placed Zeros

This design would be useful, for instance, to null a certain frequency and all its harmonics, assuming the overall lowpass function of the filter is also desirable. (Otherwise, notch filters can be used.)

For instance, if for a 7 kHz sample rate nulls are desired at 805 Hz, 1.61 kHz, 2.415 kHz and 3.22 kHz, this filter will do the job since 0 to π on this normalized scale maps to 0 to $F_s / 2$. The magnitude response of this filter, scaled for a 0 dB passband gain, is shown as the solid line in **Figure 142**. Now, if rather than true zeros, which drive the response to $-\infty$, a shallower depression in the response is desired, the zeros can be moved off the unit circle and toward its center. For instance, the dotted line in **Figure 142** shows the result of repeating this exercise with the magnitude of 0.9 rather than 1.

The coefficients for the magnitude of 0.9 case are

1.0000 1.1794 0.9669 0.9256 0.7915 0.7497 0.6344 0.6268 0.4305

The symmetry is lost, which is to be expected since **Equation 36** is not symmetric unless $r = 1$.

Chapter 4 FIR Filters

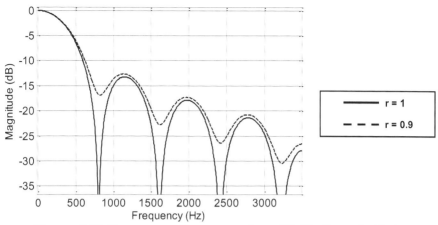

Figure 142 Frequency Response for a 7 kHz Sample Rate

It is interesting to compare the z-plane plots and group delays of the two filters (**Figure 143**). Note the difference in the magnitude of the zeros. Further note that the group delay, a constant 4 in the case of $r = 1$, is no longer constant for magnitude 0.9. We expected this since we know the filter is no longer symmetric. However, it is also worth noting that the group delay is nearly constant in the passband, which may still be satisfactory for many applications. Of course there are also applications where group delay is not a major concern.

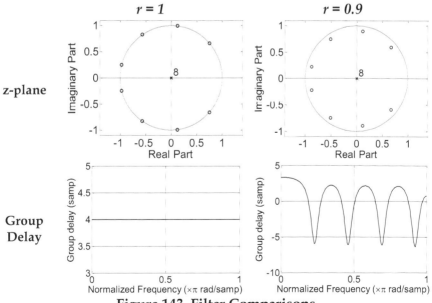

Figure 143 Filter Comparisons

165

Digital Filters for Everyone

Chapter 5: Two Dimensional Filters

The two-dimensional filters we discuss are all FIR filters. As such, all of the properties of FIR filters apply: the impulse responses are just the filter coefficients; group delay and frequency response all have the same meaning, though of course all of these concepts are generalized to two dimensions.

Two-dimensional filters are applied using two-dimensional convolution – flipping one of the items (both left-right and up-down) and moving it past the other, overlapping first only lower right and upper left elements, and then shifting element by element (right, then down at the left edge, and right again) until only the upper left and lower right elements overlap, as shown below:

$$\begin{bmatrix} a & b \\ c & d \end{bmatrix} \otimes \begin{bmatrix} e & f & g \\ h & i & j \\ k & l & m \end{bmatrix} \Rightarrow \begin{bmatrix} d & c \\ b & a \end{bmatrix} \rightarrow$$

$$\downarrow \begin{bmatrix} e & f & g \\ h & i & j \\ k & l & m \end{bmatrix} \Rightarrow$$

$$\begin{bmatrix} ae & be+af & bf+ag & bg \\ ce+ah & de+cf+bh+ai & df+cg+bi+aj & dg+bj \\ ch+ak & dh+ci+bk+al & di+cj+bl+am & dj+bm \\ ck & dk+cl & dl+cm & dm \end{bmatrix}$$

Notice that the dimension of the result is larger than that of the constituent parts. An n x n entity convolved with one of size m x m results in dimension (n + m – 1) x (n + m – 1).

The formal definition of 2D convolution is shown in **Equation 37**:

$$F(n,m) \otimes G(n,m) = \sum_{j=-M}^{M} \sum_{i=-N}^{N} F(n-i, m-j) G(i,j)$$ **Equation 37**

Two-dimensional Convolution

2.5.1 Some Standard 2D Filters [Reference 11]

Many simple image processing tasks are accomplished using one of a few standard types of 2D filters. First let us consider some simple averaging filters:

$$h = \frac{1}{4}\begin{bmatrix} 1 & 1 \\ 1 & 1 \end{bmatrix} \qquad h = \frac{1}{9}\begin{bmatrix} 1 & 1 & 1 \\ 1 & 1 & 1 \\ 1 & 1 & 1 \end{bmatrix} \qquad \text{Equation 38}$$

Averaging Filters

At left in **Figure 144** is a test image. The axis numbers show the number of pixels, 29 pixels square. At right is that same image but filtered with the 2 x 2 averaging filter defined above. Notice that the size has increased, that there is now a dark ring around the border, and that the white spot in the center of the image is broader.

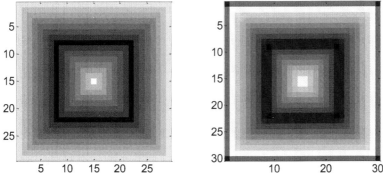

Figure 144 Test Image (L); Filtered Image (R)

To better understand what's going on, let's take a slice horizontally through the centers of these images as shown in Figure **145**. The dark region at the border of the filtered image clearly contains lower pixel values. Notice also that the image center which was previously one pixel is now two pixels wide. Furthermore, notice the smoothing which is obvious where the abrupt transitions take place.

The dark region results because when the edge of the filter overlaps the image by only one pixel, the other edge of the filter is not interacting with the image at all. (In other words, the coefficients at the outer edge of the filter are multiplied by zeros.) For this reason, these resulting values (the dark border in this case) are not valid and should be discarded. Looking back at our example convolution, only the elements involving all four terms of the filter are valid. In the case of our image convolution, this means that rather than one pixel larger, the resulting image now becomes one pixel smaller in each dimension than the original.

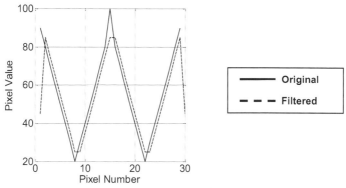

Figure 145 Slices of Images

It would make sense now to align the centers of the original and filtered images, but we have broadened our image peak into two identical values; the true center of the resulting image is actually between these two pixels. It is for this reason that most image filtering is done with odd filter sizes. This preserves the original sampling grid as shown in the next example.

The 3 x 3 averaging filter defined above would solve the sampling grid problem, but let's consider a 3 x 3 Gaussian filter, as shown in **Equation 39**, one of the true workhorses of image processing. This provides a better frequency response than does the straight average.

$$h = \frac{1}{16} \begin{bmatrix} 1 & 2 & 1 \\ 2 & 4 & 2 \\ 1 & 2 & 1 \end{bmatrix}$$ **Equation 39**

3 x 3 Gaussian Filter

Applying this filter and retaining only the valid values results in the right-hand image of **Figure 146**; the original test image is repeated at left for comparison.

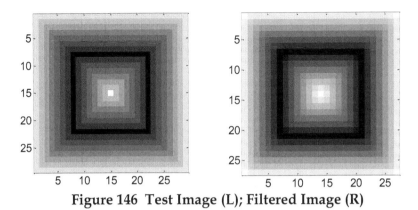
Figure 146 Test Image (L); Filtered Image (R)

Slices through the centers of these two images are shown in **Figure 147**. Notice that we achieved even more smoothing than with the 2 x 2 filter above, and that we have a unique center point as predicted.

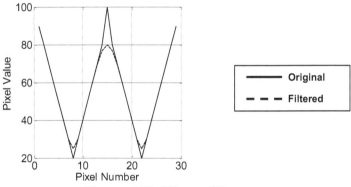
Figure 147 Slices of Images

Other common 2D filters find gradients. Some popular ones are the Roberts, Prewitt and Sobel operators, as shown below:

$$h_{Rx} = \begin{bmatrix} -1 & 0 \\ 0 & 1 \end{bmatrix} \quad h_{Ry} = \begin{bmatrix} 0 & -1 \\ 1 & 0 \end{bmatrix} \quad \text{Equation 40}$$

Roberts Operators

$$h_{Px} = \begin{bmatrix} -1 & -1 & -1 \\ 0 & 0 & 0 \\ 1 & 1 & 1 \end{bmatrix} \quad h_{Py} = \begin{bmatrix} -1 & 0 & 1 \\ -1 & 0 & 1 \\ -1 & 0 & 1 \end{bmatrix} \quad \text{Equation 41}$$

Prewitt Operators

$$h_{Sx} = \begin{bmatrix} -1 & -2 & -1 \\ 0 & 0 & 0 \\ 1 & 2 & 1 \end{bmatrix} \quad h_{Sy} = \begin{bmatrix} -1 & 0 & 1 \\ -2 & 0 & 2 \\ -1 & 0 & 1 \end{bmatrix}$$ **Equation 42**

Sobel Operators

In some respects these can be used interchangeably, though they do each have unique properties. Although some applications use the gradient phases, quite commonly we use only the magnitude, defined as follows: $M = \sqrt{G_x^2 + G_y^2}$, where G_x and G_y are the image operated upon by one of the three sets of x and y gradient operators above, and the squares of the images are computed on a point by point basis. In many cases, a threshold is used upon the magnitude to create a binary image. **Figure 148** shows Roberts and Sobel gradient magnitude images both with and without thresholding.

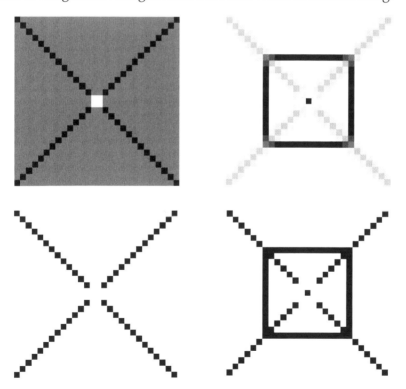

Figure 148 Gradient Magnitudes
Roberts (UL); Thresholded (LL) ; Sobel (UR); Thresholded (LR)

The Roberts image has a range of 10 to 20 and was thresholded at 12. The Sobel image has a range of 0 to 80 and was thresholded at 60.

2.5.2 Gaussian 2D Filters and Properties

It turns out that the 3 x 3 Gaussian filter of **Equation 39** is a fundamental building block of image processing due to the following excellent properties:

- The coefficients are powers of two, meaning they can be applied in hardware with simple shifts rather than multiplies.
- They have nicer frequency responses than averaging filters.
- Convolving Gaussian filters with each other results in Gaussian filters, so filters of larger sizes can be created from the 3 x 3 filter. This means, not only that the larger filters can be built and applied, but that the 3 x 3 can be applied multiple times to achieve that same effect.
- The filter is separable, meaning it can be broken down and applied as 1D filters in each dimension.

Let's discuss the last three bullet points in more detail:

Frequency response – Previously we discussed the FREQZ command in Matlab which computes frequency responses. Matlab does also have a two-dimensional version, called FREQZ2, in the Image Processing Toolbox. However, it's also a fairly straightforward matter to generalize our frequency response computation from **Chapter 1** for a 2D input filter h:

```
F = -1:0.025:1;
szh = size(h);
xD = (1 - szh(1)) / 2 : (szh(1) - 1) / 2;
yD = (1 - szh(2)) / 2 : (szh(2) - 1) / 2;
H = zeros(length(F));
for fx = 1:length(F)
    for fy = 1:length(F)
        for x = 1:szh(1)
            H(fx, fy) = H(fx, fy) + sum(h(x, :) .* ...
            exp(i .* pi .* xD(x) .* F(fx)) .* ...
            exp(i .* pi .* yD .* F(fy)));
        end
    end
end
```

This is, of course, generic Matlab code that doesn't require a special toolbox and can be used as pseudo code for any mathematical computation tool. We'll see examples of two-dimensional frequency responses below.

The frequency responses of the 3 x 3 averaging filter and the 3 x 3 Gaussian filter are shown side by side in **Figure 149**. Notice the large ripples in the response of the averaging filter. This is what we expect to see based upon our experience with 1D filters. Notice how nice and smooth are the tails of the Gaussian.

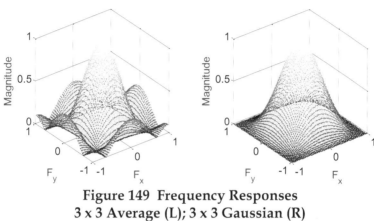

**Figure 149 Frequency Responses
3 x 3 Average (L); 3 x 3 Gaussian (R)**

Higher orders – To make larger Gaussian filters, we need only convolve the 3 x 3 until we get the filter we want. Note that, though they do remain nice integers, the coefficients do not remain powers of two. Furthermore, the larger filters become increasingly costly to implement in real-time applications. However, one need not pre-convolve the filter and then convolve that with the image. Rather, the image can be convolved multiple times with the 3 x 3 for the same effect. In some cases, this would be the more efficient way to implement a larger filter. Usually, however, the most efficient way will probably be to exploit the separability of the filter. Gaussian filters up to size 11 x 11 are tabled below.

The impulse and frequency responses of these filters are shown in **Figure 150** below. Notice that the broader the impulse response, the narrower the frequency response. This is the nature of the Fourier transform as we have encountered before.

$$G_3 = \frac{1}{16}\begin{bmatrix} 1 & 2 & 1 \\ 2 & 4 & 2 \\ 1 & 2 & 1 \end{bmatrix}$$

$$G_5 = \frac{1}{256}\begin{bmatrix} 1 & 4 & 6 & 4 & 1 \\ 4 & 16 & 24 & 16 & 4 \\ 6 & 24 & 36 & 24 & 6 \\ 4 & 16 & 24 & 16 & 4 \\ 1 & 4 & 6 & 4 & 1 \end{bmatrix}$$

$$G_7 = \frac{1}{4096}\begin{bmatrix} 1 & 6 & 15 & 20 & 15 & 6 & 1 \\ 6 & 36 & 90 & 120 & 90 & 26 & 6 \\ 15 & 90 & 225 & 300 & 225 & 90 & 15 \\ 20 & 120 & 300 & 400 & 300 & 120 & 20 \\ 15 & 90 & 225 & 300 & 225 & 90 & 15 \\ 6 & 36 & 90 & 120 & 90 & 36 & 6 \\ 1 & 6 & 15 & 20 & 15 & 6 & 1 \end{bmatrix}$$

Equation 43

$$G_9 = \frac{1}{65536} \cdot$$
$$\begin{bmatrix} 1 & 8 & 28 & 56 & 70 & 56 & 28 & 8 & 1 \\ 8 & 64 & 224 & 448 & 560 & 448 & 224 & 64 & 8 \\ 28 & 224 & 784 & 1568 & 1960 & 1568 & 784 & 224 & 28 \\ 56 & 448 & 1568 & 3136 & 3920 & 3136 & 1568 & 448 & 56 \\ 70 & 560 & 1960 & 3920 & 4900 & 3920 & 1960 & 560 & 70 \\ 56 & 448 & 1568 & 3136 & 3920 & 3136 & 1568 & 448 & 56 \\ 28 & 224 & 784 & 1568 & 1960 & 1568 & 784 & 224 & 28 \\ 8 & 64 & 224 & 448 & 560 & 448 & 224 & 64 & 8 \\ 1 & 8 & 28 & 56 & 70 & 56 & 28 & 8 & 1 \end{bmatrix}$$

Definitions for Gaussian Filters, G_n

Definitions for G_n (Equation 43) Continued

$$G_{11} = \frac{1}{1048576} \cdot$$

$$\begin{bmatrix}
1 & 10 & 45 & 120 & 210 & 252 & 210 & 120 & 45 & 10 & 1 \\
10 & 100 & 450 & 1200 & 2100 & 2520 & 2100 & 1200 & 450 & 100 & 10 \\
45 & 450 & 2025 & 5400 & 9450 & 11340 & 9450 & 5400 & 2025 & 450 & 45 \\
120 & 1200 & 5400 & 14400 & 25200 & 30240 & 25200 & 14400 & 5400 & 1200 & 120 \\
210 & 2100 & 9450 & 25200 & 44100 & 52920 & 44100 & 25200 & 9450 & 2100 & 210 \\
252 & 2520 & 11340 & 30240 & 52920 & 63504 & 52920 & 30240 & 11340 & 2520 & 252 \\
210 & 2100 & 9450 & 25200 & 44100 & 52920 & 44100 & 14400 & 5400 & 2100 & 210 \\
120 & 1200 & 5400 & 14400 & 25200 & 30240 & 25200 & 14400 & 5400 & 1200 & 120 \\
45 & 450 & 2025 & 5400 & 9450 & 11340 & 9450 & 5400 & 2025 & 450 & 45 \\
10 & 100 & 450 & 1200 & 2100 & 2520 & 2100 & 1200 & 450 & 100 & 10 \\
1 & 10 & 45 & 120 & 210 & 252 & 210 & 120 & 45 & 10 & 1
\end{bmatrix}$$

Separability – Let's put G_7 through a little bit of arithmetic:

$$\frac{1}{64}\sum_{c=1}^{7} G_7 = \begin{bmatrix} 1 & 6 & 15 & 20 & 15 & 6 & 1 \end{bmatrix}$$

$$\begin{bmatrix} 1 & 6 & 15 & 20 & 15 & 6 & 1 \end{bmatrix}^T \cdot \begin{bmatrix} 1 & 6 & 15 & 20 & 15 & 6 & 1 \end{bmatrix} =$$

$$\begin{bmatrix}
1 & 6 & 15 & 20 & 15 & 6 & 1 \\
6 & 36 & 90 & 120 & 90 & 26 & 6 \\
15 & 90 & 225 & 300 & 225 & 90 & 15 \\
20 & 120 & 300 & 400 & 300 & 120 & 20 \\
15 & 90 & 225 & 300 & 225 & 90 & 15 \\
6 & 36 & 90 & 120 & 90 & 36 & 6 \\
1 & 6 & 15 & 20 & 15 & 6 & 1
\end{bmatrix}$$

Did you follow what happened there? We summed the columns of G_7 and normalized them back to an integer value for the first element by dividing by 64. Then we formed what is known as the *outer product* and multiplied the resultant vector back out into a matrix. Lo and behold we got G_7 back again. (Note also that $64^2 = 4096$, so even if we carry the normalization factors along, the whole things still works out.)

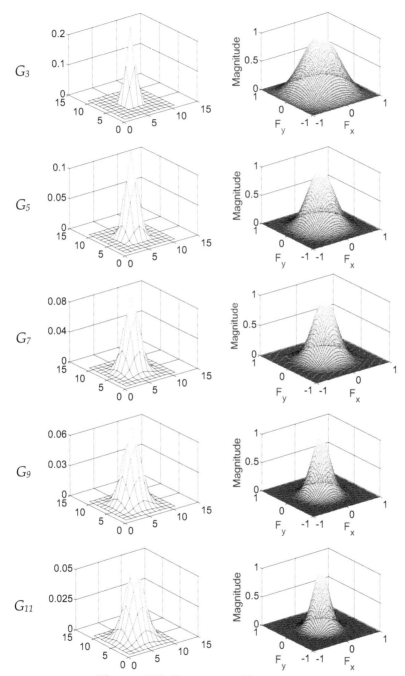

Figure 150 Frequency Responses
Impulse and Frequency Responses of Gaussian Filters

Chapter 5 Two Dimensional Filters

Now, if this is your first exposure to two-dimensional filters, you might think this is a rather ordinary occurrence. But it is not. In fact, this is one of the many wonderful properties of Gaussian filters. Let's try a non-Gaussian filter:

$$h = \frac{1}{457.4369} \begin{bmatrix} 1 & 6 & 15 & 20 & 15 & 6 & 1 \\ 6 & 16.0841 & 10.3362 & 0.5043 & 10.3362 & 16.0841 & 6 \\ 15 & 10.3362 & 1.5398 & 12.4071 & 1.5398 & 10.3362 & 15 \\ 20 & 0.5043 & 12.4071 & 0.6107 & 12.4071 & 0.5043 & 20 \\ 15 & 10.3362 & 1.5398 & 12.4071 & 1.5398 & 10.3362 & 15 \\ 6 & 16.0841 & 10.3362 & 120 & 10.3362 & 16.0841 & 6 \\ 1 & 6 & 15 & 20 & 15 & 6 & 1 \end{bmatrix}$$

$$\frac{1}{64}\sum_{c=1}^{7} h = \begin{bmatrix} 1 & 1.0210 & 1.0337 & 1.0380 & 1.0337 & 1.0210 & 1 \end{bmatrix}$$

$$\begin{bmatrix} 1 & 1.0210 & 1.0337 & 1.0380 & 1.0337 & 1.0210 & 1 \end{bmatrix}^T \cdot$$
$$\begin{bmatrix} 1 & 1.0210 & 1.0337 & 1.0380 & 1.0337 & 1.0210 & 1 \end{bmatrix} =$$
$$\begin{bmatrix} 1 & 1.0210 & 1.0337 & 1.0337 & 1.0337 & 1.0210 & 1 \\ 1.0210 & 1.0425 & 1.0555 & 1.0598 & 1.0555 & 1.0425 & 1.0210 \\ 1.0337 & 1.0555 & 1.0686 & 1.0730 & 1.0686 & 1.0555 & 1.0337 \\ 1.0337 & 1.0598 & 1.0730 & 1.0775 & 1.0730 & 1.0598 & 1.0337 \\ 1.0337 & 1.0555 & 1.0686 & 1.0730 & 1.0686 & 1.0555 & 1.0337 \\ 1.0210 & 1.0425 & 1.0555 & 1.0598 & 1.0555 & 1.0425 & 1.0210 \\ 1 & 1.0210 & 1.0337 & 1.0337 & 1.0337 & 1.0210 & 1 \end{bmatrix}$$

We notice right away that the numbers are not as nice as the Gaussian filters, meaning the scaling doesn't look as nice either. These, however, are not generally huge problems. But notice that the same operation does not lead back to the same filter. In other words, this filter is not separable as the Gaussian filters are. Indeed, the new filter has a different frequency response (**Figure 151**), as is expected due to the fact that it has different coefficients.

This filter arguably doesn't have a particularly nice frequency response to begin with; we usually don't like ripples if they can be avoided. But it serves well to illustrate that the new filter is no longer circularly symmetric since the filter was not separable.

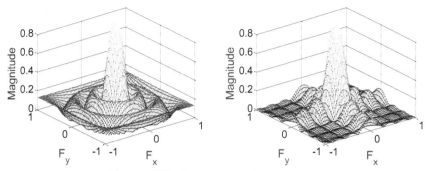

**Figure 151 Frequency Responses
Separation Attempt: Before (L); After (R)**

If the filters are applied as two-dimensional convolutions, the separability is not an issue. However, separability does allow more efficient implementation since these filters can be applied as two one-dimensional filtering operations, one in the horizontal direction, and one in the vertical.

The flip side of this separability argument is that general 2D filters cannot be created simply by multiplying out the 1D filters, as can the Gaussian filters. This means that for those cases – and there are many – where we wish to apply a circularly symmetric 2D filter that is not Gaussian, we need a different method of creating the filter. This is the subject of the next section.

2.5.3 Creating Symmetrical 2D Filters from 1D FIR Filters

There are times when we need a 2D filter different than the standard types discussed above. Typically these will have the form of odd-sized symmetrical filters; as discussed above, the odd size allows a unique center point which also retains the original sampling grid in the post filtered data. Fortunately, McClellan [**Reference 18**], Lim [**Reference 16**, pp 218-237] and others have worked out a method for creating symmetrical 2D filters from symmetrical 1D filters. Even more fortunately, the Matlab Image Processing Toolbox has the FTRANS2 to implement this. Unfortunately, though, not everyone has access to Matlab. Others just want to know what's happening under the hood, or need to implement the functionality in an embedded system. For these folks, I have written this section.

To begin, let's discuss the general theory to show you what's happening and give you the tools to extend it as needed. Then, in keeping with the spirit of this book, I provide below simple equations for filters up to 11 by 11.

First, recall that a symmetrical 1D FIR filter of size 2N+1 is described by a vector B = [b_0, b_1, ... b_{N-2}, b_{N-1}, b_N, b_{N-1}, b_{N-2}, ... b_1, b_0]. From the referenced literature we learn of transformation matrices, the first two of which are shown below. Note that the first one is trivial; the second turns out to be the basis for all the others – more on that later.

$$T_1 = 1$$

$$T_3 = \frac{1}{8}\begin{bmatrix} 1 & 2 & 1 \\ 2 & -4 & 2 \\ 1 & 2 & 1 \end{bmatrix}$$

Now, to use these transformation matrices to create a 2D filter from a 1D filter, all one needs is the following equation: $h_{3\times3} = b_1 T_1 + 2b_0 T_3$, where h_{3x3} is the 3 by 3 two dimensional filter and b_0 and b_1 are the two unique coefficients of a 1D FIR filter described by [b_0, b_1, b_0]. In other words, for the filter B = [0.5 -1 0.5], the 3 by 3 two-dimensional filter becomes $h_{3\times3} = -1 \cdot T_1 + 2 \cdot 0.5 \cdot T_3$. Notice that there is a dimensional mismatch here which can be resolved as follows:

$$h_{3\times3} = \begin{bmatrix} 0 & 0 & 0 \\ 0 & b_1 T_1 & 0 \\ 0 & 0 & 0 \end{bmatrix} + 2b_0 T_3$$

In this section we will always take the sum of items with different dimensions to mean the summation of the smaller item symmetrically about the center of the larger item. In this case the larger is 3 by 3 and the smaller is 1 by 1. So the 1 by 1 matrix is added to the center element (only) of the 3 by 3:

$$h_{3\times 3} = -1 \cdot T_1 + 2 \cdot 0.5 \cdot T_3$$

$$= -1 + 1 \cdot \frac{1}{8}\begin{bmatrix} 1 & 2 & 1 \\ 2 & -4 & 2 \\ 1 & 2 & 1 \end{bmatrix} = \begin{bmatrix} \frac{1}{8} & \frac{2}{8} & \frac{1}{8} \\ \frac{2}{8} & \frac{-4}{8}-1 & \frac{2}{8} \\ \frac{1}{8} & \frac{2}{8} & \frac{1}{8} \end{bmatrix} = \begin{bmatrix} \frac{1}{8} & \frac{2}{8} & \frac{1}{8} \\ \frac{2}{8} & \frac{-4}{8}-\frac{8}{8} & \frac{2}{8} \\ \frac{1}{8} & \frac{2}{8} & \frac{1}{8} \end{bmatrix}$$

$$= \frac{1}{8}\begin{bmatrix} 1 & 2 & 1 \\ 2 & -12 & 2 \\ 1 & 2 & 1 \end{bmatrix} = \begin{bmatrix} \frac{1}{8} & \frac{1}{4} & \frac{1}{8} \\ \frac{1}{4} & \frac{-3}{2} & \frac{1}{4} \\ \frac{1}{8} & \frac{1}{4} & \frac{1}{8} \end{bmatrix}$$

The equations for any sized 2D filter are shown below, with h_{3x3} repeated for completeness:

$$h_{3\times 3} = b_1 T_1 + 2b_0 T_3$$
$$h_{5\times 5} = b_2 T_1 + 2b_1 T_3 + 2b_0 T_5$$
$$h_{7\times 7} = b_3 T_1 + 2b_2 T_3 + 2b_1 T_5 + 2b_0 T_7$$
$$h_{9\times 9} = b_4 T_1 + 2b_3 T_3 + 2b_2 T_5 + 2b_1 T_7 + 2b_0 T_9$$
$$h_{11\times 11} = b_5 T_1 + 2b_4 T_3 + 2b_3 T_5 + 2b_2 T_7 + 2b_1 T_9 + 2b_0 T_{11} \quad \text{Equation 44}$$
$$h_{n\times n} = b_{\frac{n-1}{2}} T_1 + 2b_{\frac{n-3}{2}} T_3 + \cdots + 2b_1 T_{n-2} + 2b_0 T_n$$

$$h_{n\times n} = b_{\frac{n-1}{2}} T_1 + 2\sum_{i=0}^{\frac{n-3}{2}} b_i T_{n-2i}$$

Equations for 2D Filters from 1D

Having these equations is excellent, but notice that to put them to use we will also need definitions for higher orders of T.

To begin with we define $T_5 = 2T_3 \otimes T_3 - T_1$, where the encircled x is the convolution symbol, and where we resolve the dimension mismatch as described above (sum or difference symmetrically about the center of the larger entities).

We discussed two-dimensional convolution above. Since the Ts are symmetrical we need not concern ourselves with the flip. That is, due to symmetry they are the same when flipped.

The operation is as follows:

$$\frac{1}{8}\begin{bmatrix} 1 & 2 & 1 \\ 2 & -4 & 2 \\ 1 & 2 & 1 \end{bmatrix} \otimes \frac{1}{8}\begin{bmatrix} 1 & 2 & 1 \\ 2 & -4 & 2 \\ 1 & 2 & 1 \end{bmatrix} \Rightarrow$$

$$\frac{1}{8}\begin{bmatrix} 1 & 2 & 1 \\ 2 & -4 & 2 \\ 1 & 2 & 1 \end{bmatrix} \rightarrow$$

$$\downarrow \quad \frac{1}{8}\begin{bmatrix} 1 & 2 & 1 \\ 2 & -4 & 2 \\ 1 & 2 & 1 \end{bmatrix}$$

$$\frac{1}{64}\begin{bmatrix} 1 & 4 & 6 & 4 & 1 \\ 4 & 0 & -8 & 0 & 4 \\ 6 & -8 & 36 & -8 & 6 \\ 4 & 0 & -8 & 0 & 4 \\ 1 & 4 & 6 & 4 & 1 \end{bmatrix}$$

With the information provided it is possible to iteratively derive all of the different values of T_n.

$$T_5 = 2T_3 \otimes T_3 - T_1$$
$$T_7 = 2T_5 \otimes T_3 - T_3$$
$$T_9 = 2T_7 \otimes T_3 - T_5 \quad \text{Equation 45}$$
$$T_{11} = 2T_9 \otimes T_3 - T_7$$
$$T_n = 2T_{n-2} \otimes T_3 - T_{n-4}$$

Equations for T_n

It might be of interest to some readers, if not necessarily of practical value, that with a substantial amount of algebra (not shown) all values of T_n can be expressed in terms only of T_3 and T_1. These equations are shown in **Equation 46**.

These can be thought of as generalized polynomials where the notation Cn refers to convolution of the entity with itself n times, and where sums and differences of smaller entities are formed symmetrically about the centers of the larger ones.

$$T_5 = 2T_3^{C2} - T_1$$
$$T_7 = 4T_3^{C3} - 3T_3$$
$$T_9 = 8T_3^{C4} - 8T_3^{C2} + T_1$$
$$T_{11} = 16T_3^{C5} - 20T_3^{C3} + 5T_3$$

Equation 46

Equations for T_n in Terms of T_3 and T_1

The reasons these might not be of practical value to many readers is that they are probably just as easily generated from **Equation 45**, and because I have also provided those used in this book in **Equation 47**.

$$T_1 = 1$$

$$T_3 = \frac{1}{8}\begin{bmatrix} 1 & 2 & 1 \\ 2 & -4 & 2 \\ 1 & 2 & 1 \end{bmatrix}$$

$$T_5 = \frac{1}{32}\begin{bmatrix} 1 & 4 & 6 & 4 & 1 \\ 4 & 0 & -8 & 0 & 4 \\ 6 & -8 & 4 & -8 & 6 \\ 4 & 0 & -8 & 0 & 4 \\ 1 & 4 & 6 & 4 & 1 \end{bmatrix}$$

Equation 47

$$T_7 = \frac{1}{128}\begin{bmatrix} 1 & 6 & 15 & 20 & 15 & 6 & 1 \\ 6 & 12 & -6 & -24 & -6 & 12 & 6 \\ 15 & -6 & -15 & 12 & -15 & -6 & 15 \\ 20 & -24 & 12 & -16 & 12 & -24 & 20 \\ 15 & -6 & -15 & 12 & -15 & -6 & 15 \\ 6 & 12 & -6 & -24 & -6 & 12 & 6 \\ 1 & 6 & 15 & 20 & 15 & 6 & 1 \end{bmatrix}$$

Definitions for T_n

Equation 47 Continued: Definitions for T_n

$$T_9 = \frac{1}{512}\begin{bmatrix} 1 & 8 & 28 & 56 & 70 & 56 & 28 & 8 & 1 \\ 8 & 32 & 32 & -32 & -80 & -32 & 32 & 32 & 8 \\ 28 & 32 & -48 & -32 & 40 & -32 & -48 & 32 & 28 \\ 56 & -32 & -32 & 32 & -48 & 32 & -32 & -32 & 56 \\ 70 & -80 & 40 & -48 & 36 & -48 & 40 & -80 & 70 \\ 56 & -32 & -32 & 32 & -48 & 32 & -32 & -32 & 56 \\ 28 & 32 & -48 & -32 & 40 & -32 & -48 & 32 & 28 \\ 8 & 32 & 32 & -32 & -80 & -32 & 32 & 32 & 8 \\ 1 & 8 & 28 & 56 & 70 & 56 & 28 & 8 & 1 \end{bmatrix}$$

$$T_{11} = \frac{1}{2048} \cdot$$

$$\begin{bmatrix} 1 & 10 & 45 & 120 & 210 & 252 & 210 & 120 & 45 & 10 & 1 \\ 10 & 60 & 130 & 80 & -140 & -280 & -140 & 80 & 130 & 60 & 10 \\ 45 & 130 & 25 & -200 & -70 & 140 & -70 & -200 & 25 & 130 & 45 \\ 120 & 80 & -200 & 0 & 80 & -160 & 80 & 0 & -200 & 80 & 120 \\ 210 & -140 & -70 & 80 & -140 & 120 & -140 & 80 & -70 & -140 & 210 \\ 252 & -280 & 140 & -160 & 120 & -144 & 120 & -160 & 140 & -280 & 252 \\ 210 & -140 & -70 & 80 & -140 & 120 & -140 & 80 & -70 & -140 & 210 \\ 120 & 80 & -200 & 0 & 80 & -160 & 80 & 0 & -200 & 80 & 120 \\ 45 & 130 & 25 & -200 & -70 & 140 & -70 & -200 & 25 & 130 & 45 \\ 10 & 60 & 130 & 80 & -140 & -280 & -140 & 80 & 130 & 60 & 10 \\ 1 & 10 & 45 & 120 & 210 & 252 & 210 & 120 & 45 & 10 & 1 \end{bmatrix}$$

With the information provided it is possible to compute the circularly symmetric 2D filters from 1D for any size. To simplify that, below are presented the simplest possible equations for 2D coefficients for sizes up to 11 by 11. Due to symmetry, there are few unique coefficients. The 3 by 3 case follows:

$$h = \begin{bmatrix} h_{00} & h_{01} & h_{00} \\ h_{01} & h_{11} & h_{01} \\ h_{00} & h_{01} & h_{00} \end{bmatrix}$$ **Equation 48**

$$h_{00} = \frac{b_0}{4} \quad h_{01} = \frac{b_0}{2} \quad h_{11} = b_1 - b_0$$

Equations for h_{3x3} in Terms of b

Recall our example above where B = [0.5 -1 0.5]. According to **Equation 48**, $h_{00} = b_0 / 4 = 0.5 / 4 = 1 / 8$; $h_{01} = b_0 / 2 = 0.5 / 2 = 1 / 4$; $h_{11} = b_1 - b_0 = -1 - 0.5 = -3 / 2$. These three are the only unique coefficients, so our job is finished. Comparing to the example above we see that these equations returned the same results.

Equations for the 5 by 5, 7 by 7 and 11 by 11 cases are found below:

$$h = \begin{bmatrix} h_{00} & h_{01} & h_{02} & h_{01} & h_{00} \\ h_{01} & h_{11} & h_{12} & h_{11} & h_{01} \\ h_{02} & h_{12} & h_{22} & h_{12} & h_{02} \\ h_{01} & h_{11} & h_{12} & h_{11} & h_{01} \\ h_{00} & h_{01} & h_{02} & h_{01} & h_{00} \end{bmatrix}$$

Equation 49

$$h_{00} = \frac{b_0}{16} \quad h_{01} = \frac{b_0}{4} \quad h_{02} = \frac{3b_0}{8} \quad h_{11} = \frac{b_1}{4}$$

$$h_{12} = \frac{b_1 - b_0}{2} \quad h_{22} = b_2 - b_1 + \frac{b_0}{4}$$

Equations for h_{5x5} in Terms of b

$$h = \begin{bmatrix} h_{00} & h_{01} & h_{02} & h_{03} & h_{02} & h_{01} & h_{00} \\ h_{01} & h_{11} & h_{12} & h_{13} & h_{12} & h_{11} & h_{01} \\ h_{02} & h_{12} & h_{22} & h_{23} & h_{22} & h_{12} & h_{02} \\ h_{03} & h_{13} & h_{23} & h_{33} & h_{23} & h_{13} & h_{03} \\ h_{02} & h_{12} & h_{22} & h_{23} & h_{22} & h_{12} & h_{02} \\ h_{01} & h_{11} & h_{12} & h_{13} & h_{12} & h_{11} & h_{01} \\ h_{00} & h_{01} & h_{02} & h_{03} & h_{02} & h_{01} & h_{00} \end{bmatrix}$$

$$h_{00} = \frac{b_0}{64} \quad h_{01} = \frac{3b_0}{32} \quad h_{02} = \frac{15b_0}{64} \quad h_{03} = \frac{5b_0}{16}$$

$$h_{11} = \frac{b_1 + 3b_0}{16} \quad h_{12} = \frac{b_1}{4} - \frac{3b_0}{32} \quad h_{13} = \frac{3(b_1 - b_0)}{8}$$

$$h_{22} = \frac{b_2}{4} - \frac{15b_0}{64} \quad h_{23} = \frac{b_2 - b_1}{2} + \frac{3b_0}{16} \quad h_{33} = b_3 - b_2 + \frac{b_1 - b_0}{4}$$

Equation 50: Equations for h_{7x7} in Terms of b

$$h = \begin{bmatrix} h_{00} & h_{01} & h_{02} & h_{03} & h_{04} & h_{03} & h_{02} & h_{01} & h_{00} \\ h_{01} & h_{11} & h_{12} & h_{13} & h_{14} & h_{13} & h_{12} & h_{11} & h_{01} \\ h_{02} & h_{12} & h_{22} & h_{23} & h_{24} & h_{23} & h_{22} & h_{12} & h_{02} \\ h_{03} & h_{13} & h_{23} & h_{33} & h_{34} & h_{33} & h_{23} & h_{13} & h_{03} \\ h_{04} & h_{14} & h_{24} & h_{34} & h_{44} & h_{34} & h_{24} & h_{14} & h_{04} \\ h_{03} & h_{13} & h_{23} & h_{33} & h_{34} & h_{33} & h_{23} & h_{13} & h_{03} \\ h_{02} & h_{12} & h_{22} & h_{23} & h_{24} & h_{23} & h_{22} & h_{12} & h_{02} \\ h_{01} & h_{11} & h_{12} & h_{13} & h_{14} & h_{13} & h_{12} & h_{11} & h_{01} \\ h_{00} & h_{01} & h_{02} & h_{03} & h_{04} & h_{03} & h_{02} & h_{01} & h_{00} \end{bmatrix}$$

$$h_{00} = \frac{b_0}{256} \quad h_{01} = \frac{b_0}{32} \quad h_{02} = \frac{7b_0}{64} \quad h_{03} = \frac{7b_0}{32} \quad h_{04} = \frac{35b_0}{128}$$

$$h_{11} = \frac{b_1}{64} + \frac{b_0}{8} \quad h_{12} = \frac{3b_1}{32} + \frac{b_0}{8} \quad h_{13} = \frac{15b_1}{64} - \frac{b_0}{8} \quad h_{14} = \frac{5(b_1 - b_0)}{16}$$

$$h_{22} = \frac{b_2 + 3(b_1 - b_0)}{16} \quad h_{23} = \frac{b_2}{4} - \frac{3b_1}{32} - \frac{b_0}{8} \quad h_{24} = \frac{3(b_2 - b_1)}{8} + \frac{5b_0}{32}$$

$$h_{33} = \frac{b_3}{4} - \frac{15b_1}{64} + \frac{b_0}{8} \quad h_{34} = \frac{b_3 - b_2}{2} + \frac{3(b_1 - b_0)}{16}$$

$$h_{44} = b_4 - b_3 + \frac{b_2 - b_1}{4} + \frac{9b_0}{64}$$

Equation 51 Equations for $h_{9 \times 9}$ in Terms of b

$$h = \begin{bmatrix} h_{00} & h_{01} & h_{02} & h_{03} & h_{04} & h_{05} & h_{04} & h_{03} & h_{02} & h_{01} & h_{00} \\ h_{01} & h_{11} & h_{12} & h_{13} & h_{14} & h_{15} & h_{14} & h_{13} & h_{12} & h_{11} & h_{01} \\ h_{02} & h_{12} & h_{22} & h_{23} & h_{24} & h_{25} & h_{24} & h_{23} & h_{22} & h_{12} & h_{02} \\ h_{03} & h_{13} & h_{23} & h_{33} & h_{34} & h_{35} & h_{34} & h_{33} & h_{23} & h_{13} & h_{03} \\ h_{04} & h_{14} & h_{24} & h_{34} & h_{44} & h_{45} & h_{44} & h_{34} & h_{24} & h_{14} & h_{04} \\ h_{05} & h_{15} & h_{25} & h_{35} & h_{45} & h_{55} & h_{45} & h_{35} & h_{25} & h_{15} & h_{05} \\ h_{04} & h_{14} & h_{24} & h_{34} & h_{44} & h_{45} & h_{44} & h_{34} & h_{24} & h_{14} & h_{04} \\ h_{03} & h_{13} & h_{23} & h_{33} & h_{35} & h_{35} & h_{35} & h_{33} & h_{23} & h_{13} & h_{03} \\ h_{02} & h_{12} & h_{22} & h_{23} & h_{24} & h_{25} & h_{24} & h_{23} & h_{22} & h_{12} & h_{02} \\ h_{01} & h_{11} & h_{12} & h_{13} & h_{14} & h_{15} & h_{14} & h_{13} & h_{12} & h_{11} & h_{01} \\ h_{00} & h_{01} & h_{02} & h_{03} & h_{04} & h_{05} & h_{04} & h_{03} & h_{02} & h_{01} & h_{00} \end{bmatrix}$$

$$h_{00} = \frac{b_0}{1024} \quad h_{01} = \frac{5b_0}{512} \quad h_{02} = \frac{45b_0}{1024} \quad h_{03} = \frac{15b_0}{128} \quad h_{04} = \frac{105b_0}{512} \quad h_{05} = \frac{63b_0}{256}$$

$$h_{11} = \frac{b_1 + 15b_0}{256} \quad h_{12} = \frac{b_1}{32} + \frac{65b_0}{512} \quad h_{13} = \frac{7b_1 + 5b_0}{64} \quad h_{14} = \frac{7b_1}{32} - \frac{35b_0}{256}$$

$$h_{15} = \frac{35(b_1 - b_0)}{128} \quad h_{22} = \frac{b_2}{64} + \frac{b_1}{8} + \frac{25b_0}{1024} \quad h_{23} = \frac{3b_2}{32} + \frac{b_1}{8} - \frac{25b_0}{128}$$

$$h_{24} = \frac{15b_2}{64} - \frac{b_1}{8} - \frac{35b_0}{512} \quad h_{25} = \frac{5(b_2 - b_1)}{16} + \frac{35b_0}{256} \quad h_{33} = \frac{b_3 + 3(b_2 - b_1)}{16}$$

$$h_{34} = \frac{b_3}{4} - \frac{3b_2}{32} - \frac{b_1}{8} + \frac{5b_0}{64} \quad h_{35} = \frac{3(b_3 - b_2)}{8} + \frac{5(b_1 - b_0)}{32}$$

$$h_{44} = \frac{b_4}{4} - \frac{15b_2}{64} + \frac{b_1}{8} - \frac{35b_0}{256} \quad h_{45} = \frac{b_4 - b_3}{2} + \frac{3(b_2 - b_1)}{16} + \frac{15b_0}{128}$$

$$h_{55} = b_5 - b_4 + \frac{b_3 - b_2}{4} + \frac{9(b_1 - b_0)}{64}$$

Equation 52 Equations for h_{11x11} in Terms of b

Chapter 6 Implementation, Tips and Tricks

Chapter 6: Implementation, Tips and Tricks

Other sections provide a variety of information on filter theory and design. In this one we cover some of the practical issues one might encounter when dealing with filters.

2.6.1 Nomenclature

In this book we have used a particular filter nomenclature which is commonly used, but is not the only nomenclature one is likely to encounter. For instance, **Figure 152** shows a second-order IIR filter with b_n used to denote the feedforward coefficients, and a_n used to denote the feedback coefficients. Some authors use other symbology including, unfortunately, these same symbols for opposite purposes.

Furthermore, some authors absorb into the a coefficients the negative signs, shown where the feedback paths feed into the summer in the figure. For this reason, the transfer function of **Equation 53** sometimes has negative signs in the denominator.

In addition to the differences mentioned above, one will sometimes encounter a denominator that contains an a_0 term. This situation is remedied by dividing all numerator and denominator coefficients by a_0 and, thereby, achieving the form of **Equation 53** and the exact filter function intended.

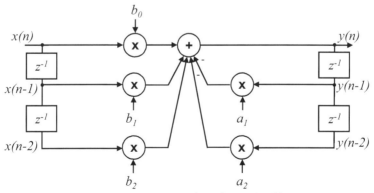

Figure 152 Second-order IIR Filter

$$H(z) = \frac{b_0 + b_1 z^{-1} + b_2 z^{-2}}{1 + a_1 z^{-1} + a_2 z^{-2}}$$ Equation 53

Transfer Function of the 2nd-order IIR Filter of Figure 11

2.6.2 Software Implementation

Though digital designers will surely implement filters from the block diagrams of **Section 1.1.6** those implementing filters in software might find useful generic Matlab code for various filter types. These could be executed in Matlab, or, equivalently, the FILTER command (Y = FILTER(B, [A], X) could be used. However, these should also be useful as pseudo code for those implementing in other languages.

```
function Y = FIR_imp(B, X)
%
% Implements FIR (Figure 12)
%
% Usage: Y = FIR_imp(B, X)
%   - B numerator coefficients
%   - X input vector
%   - Y output vector
%
states = zeros(1, length(B) - 1);
for i = 1:length(X)
   Y(i) = [X(i) states] * B';
   states(2:end) = states(1:end-1);
   states(1) = X(i);
end

function Y = DFI_imp(B, A, X)
%
% Implements DFI (Figure 13)
%
% Usage: Y = DFI_imp(B, A, X)
%   - B numerator coefficients
%   - A denominator coefficients (size of B)
%   - X input vector
%   - Y output vector
```

```
%
if length(B) ~= length(A)
   error('Size Mismatch!')
end
if A(1) ~= 1
   A = A ./ A(1);
end
Xstates = zeros(1, length(B) - 1);
Ystates = Xstates;
for i = 1:length(X)
   Y(i) = [X(i) Xstates] * B' - Ystates * A(2:end)';
   Xstates(2:end) = Xstates(1:end-1);
   Xstates(1) = X(i);
   Ystates(2:end) = Ystates(1:end-1);
   Ystates(1) = Y(i);
end
```

```
function Y = DFII_imp(B, A, X)
%
% Implements Direct Form II (Figure 14)
%    or FIR if A = 1
%
% Usage: Y = DFII_imp(B, A, X)
%  - B numerator coefficients
%  - A denominator coefficients (size of B or 1 (for FIR))
%  - X input vector
%  - Y output vector
%
lA = length(A);
lB = length(B);
if lA ~= lB & lA ~= 1
   error('A must be 1 or same size as B')
end
if A(1) ~= 1
   A = A ./ A(1);
end
states = zeros(1, lB - 1);
```

```
for i = 1:length(X)
   if lA == 1
      T = X(i);
   else
      T = [X(i) -states] * A';
   end
   Y(i) = [T states] * B';
   states(2:end) = states(1:end-1);
   states(1) = T;
end


function Y = DFIIa_imp(B, A, X)
%
% Implements Direct Form II alternate (Figure 15)
%
% Usage: Y = DFIIa_imp(B, A, X)
%    - B numerator coefficients
%    - A denominator coefficients (size of B or 1)
%    - X input vector
%    - Y output vector
%
lA = length(A);
lB = length(B);
if lA ~= lB
   error('A must be same size as B')
end
if A(1) ~= 1
   A = A ./ A(1);
end
states = zeros(1, lB - 1);
for i = 1:length(X)
   Y(i) = states(end) + X(i) * B(1);
   for j = 1:lB - 2
      states(end - j + 1) = B(j + 1) * X(i) - A(j + 1) * Y(i) + states(end - j);
   end
   states(end - lB + 2) = B(lB) * X(i) - A(lB) * Y(i);
end
```

```
function Imout = FIR2D_imp(h, Imin)
%
% Implements 2DFIR (Chapter 5: )
%
% Usage: Imout = FIR2D_imp(h, Imin)
%   - h 2D filter
%   - Imin input image
%   - Imout output image
%
% NOTE: Does not flip. Assumes filter is either symmetric or pre-flipped.
%       Filters only 'valid' region.
%
szh = size(h);
szim = size(Imin);
offi = fix(szh(1) / 2);
offj = fix(szh(2) / 2);
Imout = zeros(szim - 2 .* [offi, offj]);
for i = 1 + offi : szim(1) - offi
   for j = 1 + offj : szim(2) - offj
      Imout(i - offi, j - offj) = sum(sum(Imin(i - offi : i + offi, j - offj : j + offj) .* h));
   end
end
```

2.6.3 Fixed Point Implementation

Many filters get implemented in fixed point arithmetic, complicating the designer's design decisions, as we discuss more fully here.

In the **Theory** section there is a sub-section called **Noise Characteristics**. If this discussion is not fresh on your mind, go back and reread it before reading on. A thorough understanding is assumed in what follows.

Figure 153, repeated from the **Noise Characteristics** section, is useful in considering fixed point implementations.

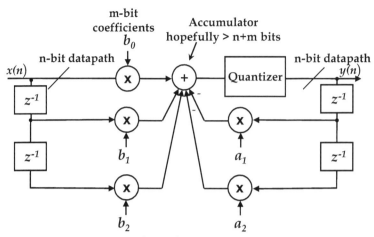

Figure 153 2nd-order Filter with Quantizer Shown

Generally speaking, I like to mentally divide up my datapath as shown in **Figure 154**.

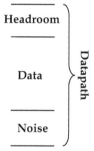

Figure 154 Datapath Model

The datapath must have room for whatever data is coming in. If any gain is to be applied in the system, there must also be headroom to accommodate it without clipping. And, if filtering or any other arithmetic is to be applied, it is wise to also allocate noise bits. The point of noise bits is not that it's desirable to carry noise along with the data, but rather that some noise will be generated by the arithmetic and subsequent re-quantization, and it's preferable to have that noise accumulate in bits beneath the data. It's very likely that there will be a final quantization before the data leaves the system, but it is preferable to do this only once to prevent accumulation of quantization noise in the lower data bits. Furthermore, some filters do amplify noise somewhat. It's also preferable to have this happen in the noise bits than in the data bits.

Now, with these concepts in mind, let us review these additional considerations for fixed point implementations:

1. What is the nature of the input signal? We know that we have an n-bit datapath, but how does the signal get to that datapath? Has anything happened to scale it such that its MSB never reaches bit n - 1? Or, is it a signal that can sometimes be highly attenuated before reaching the filter, but which still needs to be filtered and preserved properly? Unless the filter designer is also the system designer, she might not have total control over this situation, but these questions need to be resolved at the system level.

2. If the filter is to be a fixed implementation, what are the coefficients? If the coefficients will change, what is the range of possible coefficients? From the examples in this book the reader is aware that coefficients can have a fairly wide range of values, including a sizeable number before the decimal. From our **Noise Characteristics** section we are also aware that heavy quantization can change the filter or even make it go unstable.

3. What is the general effect of the filters to be implemented? If we expect that important information in the signal will be highly attenuated by the filter, but still need to be available downstream, the architecture needs to accommodate this.

Note that none of this gives us an exact answer to the question of how many bits are needed, but the thought process can help us figure that out.

Let's consider a more specific case. Years ago I worked with a client who had a fairly unique system design: there was a gain control upstream from the filters, meaning the signal could come into the system severely attenuated. Then, one branch in the system had a fairly low-frequency-cutoff lowpass filter. Notice in the **Handbook** section that it's not uncommon for a lowpass filter to have very small numerator coefficients. To further exacerbate matters, a "cost-efficient" architecture had been employed; unlike that shown in **Figure 153**, this one re-quantized the data after each multiply. This near perfect storm of system constraints combined to create a situation where the signal would sometimes cut out altogether in that one low frequency path.

I don't recall the exact parameters of that system, but let's construct a similar situation: referring to **Figure 155**, imagine that a 16-bit signal had been attenuated such that its MSB was 10 bits down. This means that what began as sixteen data bits is now effectively six. In addition, the intermediate quantizer effectively eliminates any noise bits. Now, if this signal were lowpass filtered by a second-order Butterworth with a cutoff of 100 Hz at a 48 kHz sample rate, the first coefficient applied would be around 0.00004, which would try to shift the data down by another fourteen bits. Of course, once the data is shifted by that amount, none of it remains; the signal cuts out.

Headroom
4 bits

Data
16 bits
(6 bits)

Noise
0 bits

Figure 155 Actual Datapath

So what are the solutions to such problems? To begin with, upstream attenuation must be applied with care. If it's going to happen, at least it would be good for the system designer to know about it and work accordingly. If there is always a large attenuation, the datapath model can be reworked to accommodate it. Or, if it only happens at times, is there any way to alert the downstream system to that the data can be handled differently for those cases?

Secondly, the intermediate quantizer is a bad idea. Note that, while a Butterworth lowpass has tiny numerator coefficients, it actually has a 0 dB passband. This means that even though a great attenuation is applied at the input, the output of the filter is full scale. Sufficient internal precision allows the filter to automatically recover.

It should not be ignored that a different filter might not have had such tiny input coefficients, or that there do exist other filter structures that actually use different coefficients. Such topics are worth consideration, but let us continue with the current example.

If we're willing to drop the intermediate quantizer, and if we can do nothing about the upstream attenuation, we can at least consider using our headroom. If we do not intend to apply any gain in the system, we can shift our input data up by four bits. Now we find that our ten-bit attenuation puts our MSB only six bits down. If our first coefficient will be shifting the data down by another fourteen bits, how many noise bits will we need?

Figure 156 shows another configuration. This time we have no headroom, since we've decided we don't need it. And due to the variable upstream attenuation, we know that our input signal can be attenuated between zero and ten bits. In the case where it's ten, and where we apply a filter that attenuates by another fourteen, the allocated twelve noise bits will contain at least the four MSBs internally until the filter computation pulls it back up.

Headroom
0 bits

Data
16 bits

Noise
12 bits

Figure 156 New Datapath

Is four bits enough? Again this is system dependent. Note that this signal is lowpass filtered at 100 Hz, so we are talking only about some low frequencies. If the ten-bit attenuation is not likely to happen very often, we might be able to live with this. Still, four bits is not a lot. A simulation might be a useful tool here to understand the effect. Without further information, we'd probably want to consider using a few more noise bits.

So now my readers get a glimpse of the frustration that my colleagues and clients have experienced over the years. On one hand, fixed point implementation is not rocket science. On another, dealing properly with it does require much system knowledge, careful planning, thinking through the eventualities, simulation and so forth. The great thing is that a good simulation where quantizers are carefully modeled will go a long way in understanding the performance trade offs.

2.6.4 IIR versus FIR

Sometimes there is concern over whether an IIR filter or an FIR filter should be used. For many applications it does not matter, so long as the desired transfer function can be achieved with an affordable degree of computation. This section discusses some of the tradeoffs to consider when choosing between IIR and FIR filters:

1. **Phase linearity** – as discussed throughout this book, phase linearity leads to a constant group delay, which can be easily compensated. In some applications this is useful or even necessary. Bessel IIR filters exhibit very linear phase responses in their passbands, and it is also possible to design linear phase IIR filters through other methods, such as discussed in the **Advanced Topics** chapter. However, FIR filters enjoy a wide popularity since every symmetric FIR filter has linear phase.

2. **Noise characteristics** – The **Theory** section of this book discusses the noise that arises from filter implementations. Generally speaking, this is caused by quantization and may be exacerbated by feedback. Since FIR filters do not have feedback, some feel more comfortable using them. However, it generally takes a much larger filter order to achieve a certain characteristic with an FIR than with an IIR filter. Therefore, while there is no feedback to exacerbate the noise, there is often a significant degree of quantization in an FIR implementation. Also, for many IIR filters implemented in today's high-precision architectures, the noise problem is at least manageable, if not negligible.

3. **Implementation cost** – IIR filters tend to be much lower in order than FIR, since poles used along with zeros allow creation of a variety of transfer functions much more easily. However, there are also tricks, as discussed below, to make FIR filters extremely efficient. Therefore, the implementation cost issue is not always straightforward.

Chapter 6 Implementation, Tips and Tricks

In any case, these are some of the considerations that go into choosing a filter type.

2.6.5 Parallel versus Cascade

Generally speaking filters can be implemented either by cascading sections or by combining them in parallel. The effect on the overall transfer function is different for each case: in the case of cascading, the overall transfer function is the product of the elements, while in the case of parallel combinations, it is the sum. Diagrams depicting both are shown in **Figure 157**. In either case, a convenient way to convert between the two forms is to factor the overall transfer function into poles and zeros, and then to reconstruct the new transfer function in the other form.

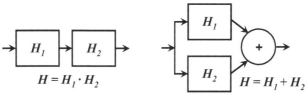

Figure 157 Parallel versus Cascade Realizations

If two second-order filters are combined in a cascade configuration, the resulting transfer function is as shown in **Equation 54**:

$$H_1(z) = \frac{b_{01} + b_{11}z^{-1} + b_{21}z^{-2}}{1 + a_{11}z^{-1} + a_{21}z^{-2}} \text{ and } H_2(z) = \frac{b_{02} + b_{12}z^{-1} + b_{22}z^{-2}}{1 + a_{12}z^{-1} + a_{22}z^{-2}},$$

$$\frac{b_{01}b_{02} + (b_{01}b_{12} + b_{11}b_{02})z^{-1} + (b_{01}b_{22} + b_{11}b_{12} + b_{21}b_{02})z^{-2} + (b_{11}b_{22} + b_{21}b_{12})z^{-3} + b_{21}b_{22}z^{-4}}{1 + (a_{11} + a_{12})z^{-1} + (a_{11}a_{12} + a_{21} + a_{22})z^{-2} + (a_{11}a_{22} + a_{12}a_{21})z^{-3} + a_{21}a_{22}z^{-4}}$$

Transfer Function:
Cascade of Two 2nd-order IIR Filters   **Equation 54**

If rather than the cascade configuration, these two are combined in parallel, the resulting transfer function is as shown in **Equation 55**. Note that this equation can be written as a single fraction, since the denominators are common. It was separated to make it fit the page.

$$\frac{b_{01}+b_{02}+(a_{11}b_{02}+a_{12}b_{01}+b_{11}+b_{12})z^{-1}+(a_{11}b_{12}+a_{12}b_{11}+a_{21}b_{02}+a_{22}b_{01}+b_{21}+b_{22})z^{-2}}{1+(a_{11}+a_{12})z^{-1}+(a_{11}a_{12}+a_{21}+a_{22})z^{-2}+(a_{11}a_{22}+a_{12}a_{21})z^{-3}+a_{21}a_{22}z^{-4}}$$

$$+\frac{(a_{11}b_{22}+a_{12}b_{21}+a_{21}b_{12}+a_{22}b_{11})z^{-3}+(a_{21}b_{22}+a_{22}b_{21})z^{-4}}{1+(a_{11}+a_{12})z^{-1}+(a_{11}a_{12}+a_{21}+a_{22})z^{-2}+(a_{11}a_{22}+a_{12}a_{21})z^{-3}+a_{21}a_{22}z^{-4}}$$

**Transfer Function:
Combination of 2nd-order Filters**   **Equation 55**

For example, beginning with two equalization filters designed using the method of the **Handbook** section: F_s = 15 kHz, f_{c1} = 100 Hz, BW_1 = 100 Hz, f_{c2} = 1 kHz, BW_2 = 1 kHz

B1 = [1.020517239018551 -1.957247176369659   0.938448282944348]
A1 = [1.000000000000000 -1.957247176369659   0.958965521962899]
B2 = [1.350592406802489 -1.506808814566801   0.298815186395021]
A2 = [1.000000000000000 -1.506808814566801   0.649407593197511]

If these two filters are cascaded, **Equation 54** gives the following coefficients:

Bc = [1.378302834029496 -4.181167545811000   4.521604471757915 -1.998917324583652 0.280422598590103]

Ac = [1 -3.464055990936459   4.557570412800192 -2.716028879458193   0.622759491577321]

Suppose that these two filters are instead combined in parallel. Using **Equation 55** we compute that the numerator coefficients of the resulting filter are:

Bp = [2.371109645821040 -7.645223536747460   9.093561260975019 -4.714946204041844   0.895988901958968]

From the equations it is clear that the denominator is the same.

Figure 158 shows the various filters of this example. The solid line is filter 1, the dashed line is filter 2, the dash-dot line is the cascade combination, and the dotted line is the parallel combination. Notice that the dash-dot line is only discernable in the region between the two original filters since it is otherwise identical to each of the individual filters. Notice also that, while the cascade combination retains the 0 dB gain outside the active regions of the filters, the parallel combination goes to 6 dB. This is due to the difference of summing or multiplying gains of 1 (0 dB).

Chapter 6 Implementation, Tips and Tricks

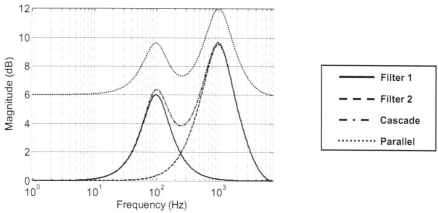

Figure 158 Parallel versus Cascade Combinations

Now, we can take one further illustrative step by computing the two second-order filters that would be necessary to create, as a cascade combination, the filter arising from the parallel combination of our original filters. This we can achieve by first factoring the resulting fourth-order filter:

Zeros: 0.966678427311648 + 0.026778526023989i
0.966678427311648 - 0.026778526023989i
0.757648417043528
0.533317800115951

Poles: 0.978623588184824 + 0.035516117624738i
0.978623588184824 - 0.035516117624738i
0.753404407283408 + 0.285988447814708i
0.753404407283408 - 0.285988447814708i

Now, two second-order filters can be created. It is best to combine the poles and zeros of similar magnitudes into each section; the first pair of zeros is combined with the first pair of poles and the second pair of (real) poles with the second pair of zeros:

B1″ = g_1 [1.000000000000000 -1.933356854623295 0.935184271285738]
A1″ = A1 = [1.000000000000000 -1.957247176369648 0.958965521962875]
B2″ = g_2 [1.000000000000000 -1.290966217159480 0.404067387038987]
A2″ = A2 = [1.000000000000000 -1.506808814566815 0.649407593197529]

Factoring does not preserve gain, so the gain must be adjusted manually, which is why gains are shown in the values of Bn″. Scaling is discussed in a subsequent section of this chapter. For now, set g_1 = g_2 = 1.539840785857768, though different values could be used so long as the product is the same. Applying these gains and using **Equation 54** we find that the resulting cascade has the following coefficients:

Bc' = [2.371109645791068 -7.645223536650825   9.093561260860080 -4.714946203982246   0.895988901947642]
Ac' = [1 -3.464055990936464   4.557570412800200 -2.716028879458198   0.622759491577322]

Other than slight differences due to the precision of the arithmetic, these are the same as Bp and Ap above.

2.6.6 Exploiting Symmetry – FIR

Although the implementation of FIR filters typically requires numerous multiplies, one of the ways some of these can be eliminated is by exploiting the symmetry of linear phase FIRs. For instance, since it is known that the latter half of the filter is a replica of the former, these coefficients need not even be stored. In addition, consider the FIR filter in **Figure 159** below:

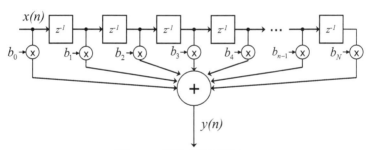

Figure 159 FIR Filter

Keeping in mind that $b_N = b_0$ and that N samples from now the sample currently being multiplied by b_0 will need to be multiplied by b_N, it is clear that this particular product could be formed once, stored, and used again at the appropriate moment.

Of course, this same analysis applies to (roughly) half of the coefficients. Therefore, for the cost of a small amount of bookkeeping overhead, half of the multiply operations, as well as half of the coefficient storage, can be saved when implementing linear phase FIR filters.

2.6.7 Polyphase Structures – FIR

FIR filters are often used in decimation and interpolation applications as shown in the non-polyphase implementation diagrams of **Figure 160**.

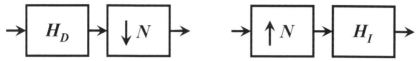

Figure 160 Decimation (L) and Interpolation (R)

In the case of decimation (left figure), a lowpass filter removes frequencies greater than the new Nyquist rate prior to decimating the signal by a factor of N, as indicated by the down arrow.

In the case of interpolation, the sample rate is increased by a factor of N by inserting zeros (typically) between samples; the following lowpass filter smoothes the signal, effectively interpolating between the original samples to create new ones.

As drawn, both of these structures are inefficient: the decimation filter runs at the original (higher) sample rate computing a number of values which are subsequently discarded by the decimation operation; the interpolation filter runs at the new (higher) sample rate often going through the motions of computing a product of zero times a coefficient.

Polyphase structures, as shown in **Figure 161**, allow the filters to run at the lower rate and avoid computing information that is not used.

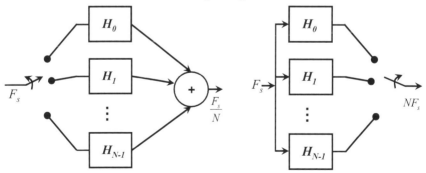

Figure 161 Polyphase Decimation (L) and Interpolation (R)

Equation 56 shows how these filters are formed. The coefficients are simply sampled from the original decimation or interpolation filters.

H_0 is the first coefficient, N^{th} coefficient, $2N^{th}$ coefficient, and so on. Likewise, H_1 is the second coefficient, $N+1^{st}$ coefficient, $2N+1^{st}$ coefficient and so on. This pattern continues until all coefficients are used. Depending on the number of coefficients and the number of branches, some branches might have fewer coefficients than others.

$$H_0 = h(0), h(N), h(2N)...$$
$$H_1 = h(1), h(N+1), h(2N+1)...$$
$$\vdots$$
$$H_{N-1} = h(N-1), h(2N-1), h(3N-1)...$$

Equation 56

Constructing Polyphase FIR Filters from Ordinary FIR Filters

2.6.8 Scaling

Many of the equations in this book return filters with 0 dB passbands. However, there are times when either the equation does not return a 0 dB passband when one is desired, or where the equation does return one and something different than 0 is desired. Generally speaking, it is a simple matter to rescale a filter using the following steps: 1) observe how the filter is scaled, and 2) multiply the numerator coefficients by a value that changes that scaling to the desired value. For example, consider the following filter, which is one of the examples from our **Parallel versus Cascade** section above:

B1 = [1.000000000000000 -1.933356854623295   0.935184271285738]
A1 = [1.000000000000000 -1.957247176369648   0.958965521962875]

Without additional numerator scaling, this filter has the frequency response shown in **Figure 162** where magnitude is shown in dB on the left, and as a linear gain on the right.

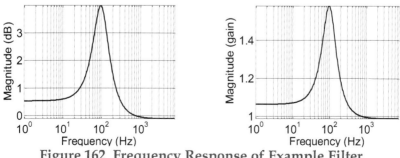

Figure 162 Frequency Response of Example Filter

Notice in this filter that there is no point that is exactly 0 dB (or, correspondingly, a gain of 1). If there is a desire to scale either the high- or low-frequency region to 0 dB, this can be accomplished by multiplying the numerator by a gain that forms a product of 1 when multiplied with the current gain. Or, if instead of scaling one of the edges, there is a desire to scale the peak from its current value of approximately 4 dB to something different, that can also be accomplished in the same way.

Zooming in on the plots, we find the values below:

	dB	Gain
LF	0.5354	1.0635
Peak	3.9878	1.5827
HF	-0.1064	0.9878

Therefore, we can scale the filter to achieve various different goals:
- 0 dB in low frequency region: $g = 1/1.0635 = 0.9403$
- 0 dB in the high frequency region: $g = 1/0.9878 = 1.0124$
- 6 dB at the peak: $g = 2/1.5827 = 1.2637$

The denominator coefficients will stay the same for each case, and the numerator coefficients for the three cases become:

B1 = $g \cdot$ [1.000000000000000 -1.933356854623295 0.935184271285738] =
[0.940300000000000 -1.817935450402284   0.879353770289980],
[1.012400000000000 -1.957330479620624   0.946780556249681], or
[1.263700000000000 -2.443183057187458   1.181792363623787]

The various cases are shown in **Figure 163**.

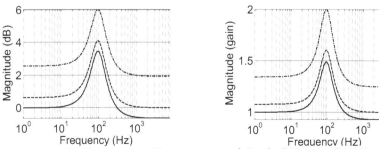

Figure 163 Frequency Responses of Scaled Example Filter: dB (L), Linear (R)

Digital Filters for Everyone

Clearly, this method can be used to scale any filter in any arbitrary way. However, for low- and high-frequency scaling, there is also another simple method that is often useful. This method arises from consideration of the frequency response given by **Equation 57** (see the **Theory** section for details).

$$H(f) = \frac{b_0 + b_1 e^{-i\frac{2\pi f}{F_s}} + b_2 e^{-i\frac{4\pi f}{F_s}} + \cdots + b_N e^{-i\frac{2N\pi f}{F_s}}}{1 + a_1 e^{-i\frac{2\pi f}{F_s}} + a_2 e^{-i\frac{4\pi f}{F_s}} + \cdots + a_N e^{-i\frac{2N\pi f}{F_s}}}$$ Equation 57

General IIR Transfer Function (Frequency Form)

Now notice what happens when f = 0 Hz (DC): the arguments of all exponential terms go to 0, making the exponentials themselves equal 1 ($e^0=1$); the response of the filter at DC is equal to the sum of the numerator coefficients divided by the sum of the denominator coefficients. (For FIR filters it is simply equal to the sum of the numerator coefficients, since the denominator is 1.)

$$H(0) = \frac{b_0 + b_1 + b_2 + \cdots + b_N}{1 + a_1 + a_2 + \cdots + a_N}$$ Equation 58

Filter Response at DC (0 Hz)

Therefore, to change the gain in the low-frequency region, we can simply divide the desired gain, g_d, by the response at 0 to find the gain to apply, g_a, as shown in **Equation 59**.

$$g_{aLF} = \frac{g_{dLF}}{H(0)} = \frac{g_{dLF}}{\frac{\sum_{k=0}^{N} b_k}{\sum_{k=0}^{N} a_k}} = \frac{g_{dLF} \sum_{k=0}^{N} a_k}{\sum_{k=0}^{N} b_k}$$ Equation 59

Modifying Gain at DC (0 Hz)

Trying this on our example filter above with a desired gain of 1 (0 dB) we get the same result as when we scaled the filter by reading the gain from the plot:

$$g_{aLF} = \frac{g_{dLF} \sum_{k=0}^{N} a_k}{\sum_{k=0}^{N} b_k} = \frac{1 \cdot [1.0 - 1.957247176369648 + 0.958965521962875]}{[1.0 - 1.933356854623295 + 0.935184271285738]}$$

$$g_{aLF} \approx 0.9403$$

Now, keeping in mind that the maximum frequency is $F_s/2$, let us try the same thing for the high-frequency scaling:

$$H\left(\frac{F_s}{2}\right) = \frac{b_0 + b_1 e^{-i\pi} + b_2 e^{-i2\pi} + \cdots + b_N e^{-iN\pi}}{1 + a_1 e^{-i\pi} + a_2 e^{-i2\pi} + \cdots + a_N e^{-iN\pi}}$$

$$H\left(\frac{F_s}{2}\right) = \frac{b_0 - b_1 + b_2 - \cdots \pm b_N}{1 - a_1 + a_2 - \cdots \pm a_N}$$

Equation 60

Filter Response at $F_s/2$ (Nyquist)

In this case, the exponential term is negative for odd numbered coefficients and positive for even numbered coefficients. Therefore, our scaling formula changes only slightly:

$$g_{aHF} = \frac{g_{dHF} \sum_{k=0}^{N} a_k (-1)^k}{\sum_{k=0}^{N} b_k (-1)^k}$$

Equation 61

Modifying Gain at $F_s/2$ (Nyquist)

Once again applying this to our example filter we get essentially the same answer as we did by reading the gains from the chart in our example above. (The sign of the center term has been reversed.)

$$g_{aHF} = \frac{g_{dHF} \sum_{k=0}^{N} a_k}{\sum_{k=0}^{N} b_k} = \frac{1 \cdot [1.0 + 1.957247176369648 + 0.958965521962875]}{[1.0 + 1.933356854623295 + 0.935184271285738]}$$

$$g_{aHF} \approx 1.0123$$

This also works for FIR filters, though the contribution of the denominator is once again identically 1.

Note in addition that this technique can technically be applied to any frequency. However, in all other cases the value of the exponential will be complex so the formulas do not work out to be quite so simple.

2.6.9 Filter Reuse

Many of the equations in this book rely upon a value γ defined as

$$\gamma = \tan\left(\pi f_c / F_s\right)$$　　　　　　　Equation 62

Intermediate Variable γ Used in Many Filter Equations

Note that γ will have the same value for every combination of f_c and F_s that achieves the same ratio:

$$\frac{f_c}{F_s} = \frac{500Hz}{10kHz} = \frac{1000Hz}{20kHz} = \frac{1900Hz}{38kHz} = \frac{500MHz}{10GHz}$$

That is, the filter will be the same as long as this ratio is the same. This property is true even for filters that do not use γ. All digital filters can be used at any sample rate and the critical frequencies will scale accordingly: if a filter that is designed for one sample rate is applied at a different rate, the corner frequency (and other critical frequencies) moves as a ratio of the sample rates, doubling if the sample rate is double, halving if it is half, and so forth.

For example, consider again the filter used in the previous section:
B1 = [1.000000000000000 -1.933356854623295 0.935184271285738]
A1 = [1.000000000000000 -1.957247176369648 0.958965521962875]

This filter was applied at a sample rate of 15 kHz, and, as seen from **Figure 162**, has a peak around 97.5 Hz when applied at that rate. Let us now plot, in **Figure 164**, the response of the same filter when applied at 10 kHz, 15 kHz (as above), 20 kHz and 25 kHz.

Notice how the original 97.5 Hz peak moves with sample rate:

$$97.5Hz \frac{10kHz}{15kHz} = 65Hz. \quad 97.5Hz \frac{20kHz}{15kHz} = 130Hz. \quad 97.5Hz \frac{25kHz}{15kHz} = 162.5Hz.$$

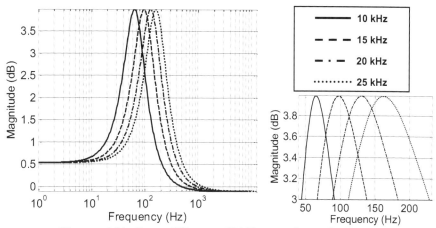

Figure 164 Same Filter at Different Sample Rates

2.6.10 Real Time Implementations via Our Simple Formulas

Although some of them are more complicated than others, the equations in this book are the simplest known ways to design digital filters, and are generally much simpler than the transform-based approach found elsewhere. This simplicity also makes the burden of a coefficient-computing real-time processor much lighter.

Many of the filter types require computation of the tangent, to form the intermediate variable γ defined in **Equation 62** above. While many modern processors will have no problem making this computation, for ultra low-cost applications, it might be desirable to avoid it. In the previous section we pointed out that γ is unique only to the ratio of f_c to F_s. Therefore, values of γ for various ratios could be pre-computed and stored to allow simple computation of coefficients. The density of this table would be a function of the precision needed for corner frequencies in the filters to be designed in real-time. Note that the values of γ could be interpolated from a sparse table, although the tangent is a highly non-linear function, so care would need to be taken to store enough points to define the curve sufficiently. Of course, since error in the value of γ only moves the corner frequency of the filter, some error would not be catastrophic for many applications.

The value of γ between the values of 0 and 0.48π is plotted in **Figure 165**. Since f_c is limited to $F_s / 2$, the argument of the tangent can technically go all the way to $\pi / 2$ making γ go to infinity. However, useful filter corners will generally be somewhat lower, and the character of the lower part of the tangent function is easier to see when plotting only to 0.48π.

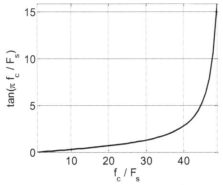

Figure 165 γ Between 0 and 0.48π

In addition to γ, some of the filter equations use trigonometric functions to precisely define constants. Such values could also be stored to save computational burden in real-time applications.

Part 3: Advanced Topics

A book that tried to tackle every aspect of digital filtering would probably never end. Therefore, while we have tried to provide the practitioner with straightforward equations that will solve many filtering problems, there are, obviously, many other topics that could be discussed. This section tackles just few of them.

Chapter 7: Linear Phase IIR Filter Design

In the **IIR** section, we have discussed Bessel filters, which have very linear phase in the passband. In addition, in the **FIR** section, we have shown that it is simple to produce exactly linear phase FIR filters. Finally, while FIR filters tend to be fairly long (high order) for a particular frequency response, we have also considered efficient implementation techniques.

This section can be thought of as another technique to reduce the computational load of FIR filters, or can be thought of as another method of producing linear phase IIR filters. As usual, there is extensive literature on this topic; however, this section covers only one approach, which was developed by Sreeram and Agathoklis in 1992 [**Reference 31**].

A discrete linear system can be expressed as follows:
$$x(n+1) = Ax(n) + bu(n)$$
$$y(n) = cx(n) + du(n)$$

Using this notation, an FIR filter, $[h_0, h_1,...,h_N]$, can be written

$$A = \begin{bmatrix} 0 & 0 & \cdots & 0 & 0 \\ 1 & 0 & \cdots & \cdots & \vdots \\ 0 & 1 & 0 & \cdots & \vdots \\ \vdots & 0 & 1 & 0 & \vdots \\ 0 & 0 & \cdots & 1 & 0 \end{bmatrix}$$

$$b = \begin{bmatrix} 1 & 0 & \cdots & 0 \end{bmatrix}^T$$
$$c = \begin{bmatrix} h_1 & h_2 & \cdots & h_N \end{bmatrix}$$
$$d = h_0$$

Try it; it really works!

To reduce this system, form the impulse-response gramian:

$$G = \sum_{k=0}^{\infty} \begin{bmatrix} h_{k+1}^2 & h_{k+1}h_{k+2} & \cdots & h_{k+1}h_{k+N} \\ h_{k+1}h_{k+2} & h_{k-2}^2 & \vdots & h_{k+2}h_{k+N} \\ \vdots & \vdots & \vdots & \vdots \\ h_{k+1}h_{k+N} & h_{k+2}h_{k+N} & \cdots & h_{k+N}^2 \end{bmatrix}$$

Compute the eigenvalues, V, and the eigenvectors, T, of G. Order the eigenvectors in decreasing eigenvalue order. ($T^T G T$ is a diagonal matrix of the eigenvalues sorted from top down, largest to smallest.)

Now compute the following values:

$$A_i = T^{-1} A T \quad c_i = cL$$
$$b_i = L^{-1} b \quad d_i = h_0$$

Partition the above system according to the order desired for the new system. If the original FIR filter has 19 taps, A_i will be 18 x 18. If you would like the new system to be order 6, make A_{i11} 6 x 6:

$$A_i = \begin{bmatrix} A_{i11} & A_{i12} \\ A_{i21} & A_{i22} \end{bmatrix} \quad \begin{matrix} c_i = \begin{bmatrix} c_{i1} & c_{i2} \end{bmatrix} \\ d_i = h_0 \end{matrix}$$

$$b_i = \begin{bmatrix} b_{i1} \\ b_{i2} \end{bmatrix}$$

Your new, reduced order system is A_{i11}, b_{i1}, c_{i1}, d_i. Turn this into a filter again using $H(z) = c(zI-A)^{-1}b+d$.

As an example, consider the 98th-order lowpass FIR filter designed using the Hamming windowed Fourier series method (see the **FIR** section) for a cutoff of 5 kHz and a sample rate of 20 kHz. The response of this filter is plotted in **Figure 166**. Of course the group delay for this filter is an exact 49 samples across the band.

Chapter 7 Linear Phase IIR Filter Design

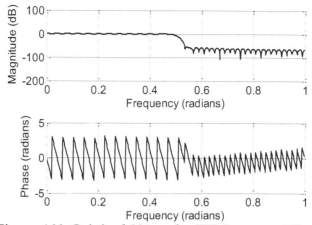

Figure 166 Original 98th-order FIR Lowpass Filter

Using the method above, we can create IIR approximations of various sizes. For example, let us begin with a 49th-order approximation as shown in **Figure 167**.

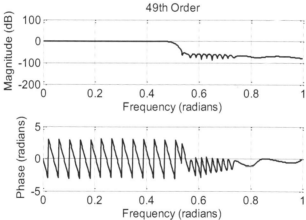

Figure 167 49th-order IIR Approximation

Note that the approximation is very good in the passband though the stopband characteristic varies in both magnitude and phase, which would typically be of little concern. Likewise, the group delay, shown in **Figure 168**, is 49 samples in the passband, though it has some ripple near the transition, and then rings in the stopband, which is also not likely a major concern since the magnitude is significantly attenuated here.

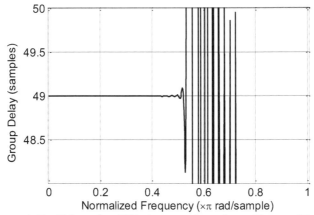

Figure 168 49th-order IIR Approximation: Group Delay

So we have approximated a linear phase FIR filter with an IIR filter of half that order. However, this is probably not a computational savings for two reasons: 1) an IIR has twice the coefficients (numerator and denominator) as the same order FIR, and 2) an IIR filter typically does not lend itself to the same kinds of implementation efficiencies as an FIR filter due to symmetry.

Now, what happens as the order of the approximation decreases? **Figure 169** shows a 29th-order approximation; the group delay follows in **Figure 170**. The stopband is very clean, but has risen to around -42.5 dB as compared to the -65 dB level of the original. In addition, the passband group delay is ringing more.

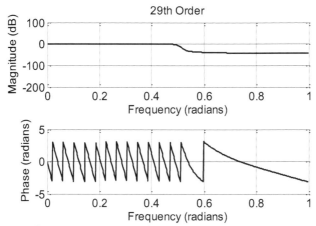

Figure 169 29th-order IIR Approximation

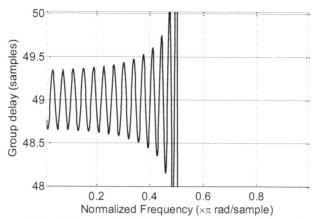

Figure 170 29th-order IIR Approximation: Group Delay

This approximation might be satisfactory for some applications. However, given its decreased stopband attenuation, one would want to compare the performance of a shorter FIR filter, which could be implemented efficiently by exploiting the FIR symmetry.

As would be expected, the approximation would continue to degrade as the order is decreased.

Digital Filters for Everyone

Chapter 8: Mapping Filters to Different Frequencies

In one of my filtering papers [**Reference 4**] I called Moorer's "cleverly-titled opus" [**Reference 20**] "the finest paper written on filter design." Dr. Moorer has also written an excellent but unpublished update to that paper which can be found on his website [**Reference 21**].

Among other gems in Moorer's papers is a description of how to use conformal mapping to move the critical frequencies of a filter. This can be accomplished by application of **Equation 63** below:

$$z^{-1} = \frac{a + z^{-1}}{1 + az^{-1}}$$

$$0 = a^2 \sin(\rho + \phi) + 2a\sin(\rho) + \sin(\rho - \phi) \qquad \textbf{Equation 63}$$

$$\rho = \frac{2\pi f_{c1}}{F_{s1}} \qquad \phi = \frac{2\pi f_{c2}}{F_{s2}}$$

Conformal Mapping (from Moorer)

Clearly, ρ is a function of the critical frequency and sample rate of the original filter; ϕ is a function of the critical frequency and sample rate of the new filter. Note also that the second equation is quadratic in a meaning that two solutions will be produced. Only one of them will be inside the unit circle; that one should be chosen.

To apply this to a second-order filter, we substitute the map equation into the transfer function and achieve the following:

$$H(z)\Big|_{\rho \to \phi} = \frac{b_0 + b_1 z^{-1} + b_2 z^{-2}}{1 + a_1 z^{-1} + a_2 z^{-2}} \Bigg|_{z^{-1} = \frac{a+z^{-1}}{1+az^{-1}}} =$$

$$\frac{b_0 + b_1 a + b_2 a^2 + \left(2b_0 a + b_1(a^2 + 1) + 2b_2 a\right) z^{-1} + \left(b_0 a^2 + b_1 a + b_2\right) z^{-2}}{1 + a_1 a + a_2 a^2 + \left(2a + a_1(a^2 + 1) + 2a_2 a\right) z^{-1} + \left(a^2 + a_1 a + a_2\right) z^{-2}}$$

Changing the Critical Frequency of a Second-order Filter   **Equation 64**

As an example let us try a second-order filter from the **Tips and Tricks** section:

B1 = [1.000000000000000 -1.933356854623295 0.935184271285738]
A1 = [1.000000000000000 -1.957247176369648 0.958965521962875]

This filter has a center frequency of 97.5 Hz at a 15 kHz sample rate. Let us use **Equation 64** to remap that to 63 Hz:

$$\rho = \frac{2\pi f_{c1}}{F_{s1}} = \frac{2\pi 97.5}{15000} = 0.040840704496667$$

$$\phi = \frac{2\pi f_{c2}}{F_{s2}} = \frac{2\pi 63}{15000} = 0.026389378290154$$

As discussed above, this results in a quadratic equation in a:
$0.067179448868211a^2 + 0.081658703957020a + 0.014450823207274 = 0$

This quadratic equation has these roots: -1.000539131887466 and -0.214991887386213. We choose the latter since it is inside the unit circle. Then, applying **Equation 64** to the original filter with this value of a returns the following coefficients (response in **Figure 171**):

B1' = [0.995744073483865 -1.948525048087132 0.953549597881191]
A1' = [1.000000000000000 -1.972564164450202 0.973286911730348]

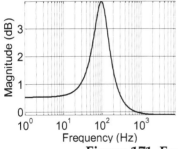 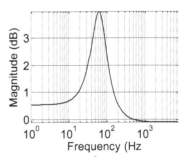

Figure 171 Frequency Responses of Original (L) and Remapped (R) Filters

From **Equation 63** it is clear that ρ and ϕ are functions of both critical frequencies and sample rates. Therefore, this technique can also be used to maintain the same critical frequency when the sample rate changes, or to change both the sample rate and the critical frequency.

Chapter 9: References

There are many good books and papers on digital filtering, though they generally approach the topic from a transform point of view: build a prototype analog filter; transform it to the desired filter type with the desired critical frequencies; transform it to digital.

Instead, we took a direct approach in this book, providing the reader with equations that deliver coefficients directly when filter parameters are input. Nevertheless, there is much to learn on the topic of filtering. I used the books and papers below while writing this book, and you will find many of them useful as well:

1. R. Allred, *Cascading IIR Filters*, Planet Analog, 14 July 2003.
2. R. Allred, *Digital Filters for Everyone*, Creative Arts & Sciences House, 2010.
3. R. Allred, *Digital Filters for Everyone, Second Edition*, CreateSpace with Creative Arts & Sciences House, 2013.
4. R. Allred, *On the Design of Higher-order Shelf Filters*, Planet Analog, 19 February 2004.
5. A. Antoniou, *Digital Filters: Analysis and Design*, McGraw-Hill, 1979.
6. N. K. Bose, *Digital Filters: Theory and Applications*, North-Holland, 1985.
7. S. M. Bozic, *Digital and Kalman Filtering, Second Edition*, Halsted Press, 1994.
8. R. Bristow-Johnson, *The Equivalence of Various Method of Computing Biquad Coefficients for Audio Parametric Equalizers*, AES 97th Convention, 3906, Nov. 94.
9. R. J. Clark et al., *Techniques for Generating Digital Equalizer Coefficients*, Journal of the Audio Engineering Society, 48(4), 281-298, 2000.
10. B. Gold and C. M. Rader, *Digital Processing of Signals*, McGraw-Hill, New York, 1969.
11. R. C. Gonzalez and R. E. Wood, *Digital Image Processing, Third Edition*, Prentice Hall, 2008.
12. R. W. Hamming, *Digital Filters*, Dover Publications, 1997.

13. Y. Hirata, *Digitalization of Conventional Analog Filters for Recording Use*, Journal of the Audio Engineering Society, 29(5):333-337, 1981.
14. E. C. Ifeachor and B. W. Jervis, *Digital Signal Processing: A Practical Approach*, Addison-Wesley, 1993.
15. L. B. Jackson, *Digital Filters and Signal Processing*, Kluwer Academic Publishers, 1986.
16. J. S. Lim, *Two Dimensional Signal and Image Processing*, Prentice Hall, 1990.
17. S. P. Lipshitz and J. Vanderkooy, *A Family of Linear-Phase Crossover Networks of High Slope Derived by Time Delay*, Journal of the Audio Engineering Society, 31(1):374-392, 1983.
18. J. H. McClellan, *The Design of Two-Dimensional Digital Filters by Transformation*, Seventh Annual Princeton Conference on Information Sciences and Systems, 247-251, Mar. 73.
19. J. McNally, *Digital Audio: Recursive Digital Filtering for High Quality Audio Signals*, BBC Research Department Report, 1981/10, 1981.
20. J. A. Moorer, *The Manifold Joys of Conformal Mapping*, Journal of the Audio Engineering Society, 31(11):826-841, 1983.
21. J. A. Moorer, *The Manifold Joys of Conformal Mapping: Applications to Digital Filtering in the Studio – 2nd Try*, Unpublished manuscript: http://www.jamminpower.com/main/unpublished.html.
22. A. V. Oppenheim and R. W. Schafer, *Digital Signal Processing*, Prentice-Hall, 1975.
23. S. J. Orfanidis, *Digital Parametric Equalizer Design with Prescribed Nyquist-Frequency Gain*, AES 101st Convention, 4361, Nov. 96.
24. T. W. Parks and C. S. Burrus, *Digital Filter Design*, Wiley-Interscience, 1987.
25. J. G. Proakis and D. G. Manolakis, *Introduction to Digital Signal Processing*, Macmillan, 1988.
26. P. A. Regalia and S. K. Mitra, *Tuneable Digital Frequency Response Equalization Filters*, IEEE Transactions on Acoustics, Speech, and Signal Processing, 35(1):118-120, 1987.
27. P. A. Regalia, S. K. Mitra and P. P. Vaidyanathan, *The Digital All-Pass Filter: A Versatile Signal Processing Building Block*, Proceedings of the IEEE, 76(1):19-37, 1988.

28. C. B. Rorabaugh, *Digital Filter Designer's Handbook, Second Edition*, McGraw-Hill, 1997.
29. Dietrich Schlichthärle, *Digital Filters: Basics and Design*, Springer, 2000.
30. J. O. Smith III, *Introduction to Digital Filters: with Audio Applications*, W3K Publishing, 2007.
31. V. Sreeram and P. Agathoklis, *Design of Linear-Phase IIR Filters via Impulse-Response Grammians*, IEEE Transactions on Signal Processing, Vol. 40, No. 2, February 1992, pp 389 - 394.
32. T. J. Terrell, *Introduction to Digital Filters*, Halsted Press, 1980.
33. D. Zaucha, *Importance of Precision on Performance for Digital Audio Filters*, Audio Engineering Society 112th Convention. Munich, May 2002.
34. U. Zolzer and T. Boltze, *Parametric Digital Filter Structures*, Audio Engineering Society 99th Convention. New York, October 1995.

Index

2D Filters, 167-186

Allpass, 6, 16-17, 49, 66-68, 122-124
Applications, 53-68

Bandpass, 6, 77-121
Bandstop, 6, 79-123
Bessel, 53, 58, 86-94, 196, 209
Butterworth, 22-32, 40, 43-49, 53, 59-60, 66-80, 86, 91-93, 94, 117-122, 194

Digital Signal, 1-2
Digital System Theory, 1-4

Cascade, 197-202
Chebychev, 53-61, 94-116
    Type I, 55-57, 95-106
    Type II, 61, 106-116
Conformal Mapping, 215-216
Cosine Window, 150-154

Equalization, 65-68, 116, 122-126

Filter Structures, 13-17
Filter Theory, 1-51
Filter Transforms, 42-49
Finite Impulse Response, FIR, 8, 9, 11-14, 18, 33, 38, 49-50, 69, 135-165, 167, 178-179, 188-189, 196, 200-204, 206, 209-213
Fourier Series Method, 135-154
Fixed Point, 37-38, 191-196
Frequency Response, 24-32, 172-173
Frequency Sampling Method, 154-162

Gibbs Phenomenon, 150-154

Group Delay, 25-32, 167
Hamming Window, 150-154
Highpass, 6, 74-162
Implementation, SW, 188-191
Impulse Response, 8-11, 32-36, 49-50, 140, 143, 167, 174
Infinite Impulse Response, IIR, 9-17, 35-36, 38-42, 49-51, 69-133, 187-196, 209-213

Linear Phase, 13, 86-94, 135-165, 196, 200, 205-213
Linear Phase IIR, 86-94, 205-213
Linkwitz-Reilly, 53, 64, 80-86, 94, 117-121
Lowpass, 6, 22, 40-49, 59, 66, 70-165, 193-195, 201, 210-211

Magnitude Response, 26-32

Noise Transfer Function, 40, 51
Notch, 6-7, 62-63, 126-127

Parallel, 197-200
Passband, 6
Phase Response, 26-32
Poles, 17-23, 37-41, 47, 50-51, 70, 129-134, 141-162, 196-199
Polyphase, 201-202

Q, 7, 37, 41, 77-79, 84-85, 91-93, 103-105, 114, 116-122
Quadratic Equation, 4, 216

Real-time, 173, 207-208
Rectangular Window, 150-154

Sample Frequency, 1
Sample Interval, 1
Sample Rate, 1

Scaling, 98, 139-142, 163, 177, 199, 202-206
Shelf, 6, 64-66, 123, 127-129
Stability, 17-24, 47, 50, 130
Step Response, 9, 32-36, 50, 56-60
Stopband, 6
SW Implementation, 188-191

Transfer Function, 9-14, 69, 129
Transforms, 42-48
Transition Band, 6
Two-dimensional Filters, 167-186, 191

Variable Q, 116-122
von Hann Window, 151-154

Window Functions, 136, 143, 150-154, 158-159, 210

Zeros, 17-26, 50-51, 129-134, 141-144, 155, 159-165, 196-201